Resources for Fire Department Occupational Safety and Health

Stephen N. Foley, *Editor*

Senior Fire Service Specialist
National Fire Protection Association

National Fire Protection Association
Quincy, Massachusetts

Product Manager: Pam Powell
Developmental Editor: Robine Andrau
Editorial-Production Services: Publishers' Design and Production Services, Inc.
Composition: Publishers' Design and Production Services, Inc.
Cover Design: Groppi Advertising Design
Manufacturing Manager: Ellen Glisker
Printer: Edwards Brothers/Lillington

Copyright © 2003
National Fire Protection Association, Inc.
One Batterymarch Park
Quincy, Massachusetts 02269

Notice Concerning Liability: Publication of this work is for the purpose of circulating information and opinion among those concerned for fire and electrical safety and related subjects. While every effort has been made to achieve a work of high quality, neither the NFPA nor the authors and contributors to this work guarantee the accuracy or completeness of or assume any liability in connection with the information and opinions contained in this work. The NFPA and the authors and contributors shall in no event be liable for any personal injury, property, or other damages of any nature whatsoever, whether special, indirect, consequential, or compensatory, directly or indirectly resulting from the publication, use of, or reliance upon this work.

This work is published with the understanding that the NFPA and the authors and contributors to this work are supplying information and opinion but are not attempting to render engineering or other professional services. If such services are required, the assistance of an appropriate professional should be sought.

NFPA No.: FDRES03
ISBN: 0-87765-486-7
Library of Congress Control No.: 2003100889

Printed in the United States of America
03   04   05   06   07   5   4   3   2   1

An Rathad Dhachaigh (The Road Home)

This guide is dedicated to the men and women of the fire service, especially my good friend Lt. Arthur Washburn (retired), who worked at NFPA for over twenty years and was dedicated to making the profession of fire fighting safer.

Contents

Foreword

Fire fighter safety is not a fad that will pass through our service and be replaced by a fascination with the latest gizmo or shiny piece of fire apparatus. Fire fighter safety is a commitment to care for our personnel over the long haul: before, during, and after the response to the emergency incident.

On the incident scene, assuring the safety of our most valuable asset, our fire fighters, is a basic responsibility of the incident commander. It is also a sacred bond and promise that bosses make to the workers—work within the command system and the incident commander will watch over you.

Fire fighter safety is a practical pursuit that pays off at the incident scene. Fire fighters and officers work more effectively if they know that their welfare is a component of the incident action plan. I have observed that all fire fighting comes to a halt when a fire fighter is in trouble—everyone's interest and efforts go toward the fire fighter's dilemma.

Before and after the incident, the bond between fire officers and fire fighters is no less important. Each fire department member has the responsibility to work together to maintain the health of our members. This is a long-term approach that shows our members that we care about them and want them to continue to be able to contribute to our organizational goals and to the quality of their lives on and off the job.

This book, *Resources for Fire Department Occupational Safety and Health*, brings together the latest information on the subject. The contents of the book identify the problems, examine the components of our response to those problems, cite examples of when our best safety efforts fail, and provide insights into the standards that exist to assure the safety and health of our members. The contents of the book, when applied as a comprehensive health and safety program, will save lives, extend the careers, and improve the quality of fire fighters' lives after they hang up their helmets.

The editor of this excellent resource, Stephen N. Foley, has been an advocate and recognized expert regarding fire fighter health and safety issues for twenty plus years. As a fire chief and member of the NFPA Technical Committee that authored the first edition of NFPA 1500 in 1987, Chief Foley brought a practical perspective to the process, and he applied the requirements of the standard to his own fire department. After he retired and began his work with NFPA, Stephen continued to work for the safety of fire fighters—this time by his able facilitation of several NFPA technical committees with scopes that impact fire fighter safety. Stephen has also been a personal friend and safety colleague for over twenty years.

I encourage the reader to make the most of this book. It's a comprehensive treatment of a subject that makes an enormous impact on our ability to serve our customers, and to survive the process to live another day.

Alan V. Brunacini
Fire Chief, Phoenix Fire Department

Acknowledgments

"Many hands make light work."

—Charlie Chan

There are some folks who worked tirelessly to get this project to completion. First let me thank my boss, Gary Tokle, for his insight and perseverance on the merits of this project. Next, thanks to the contributors: John Granito, Ed.D.; Kevin Roche, Ph.D.; Murrey Loflin; Jonathan Kipp; Chief Edward Stinette and Deputy Chief Glenn Benarick of the Fairfax County Fire and Rescue Department, Virginia; Rich Duffy and Mike Crouse of International Association of Fire Fighters; Gary Briese, Executive Director of International Association of Fire Chiefs; M. S. Bogucki, Ph.D., Yale University; Ron Sarnicki, Executive Director of National Fallen Firefighters Foundation; and Paul LeBlanc, Michael Karter, and Stephen Badger of NFPA's Fire Analysis and Research Division.

Thanks also to NFPA staff contributors: Maureen Brodoff, General Counsel; Robert Duval, Chief Fire Investigator; Laurence Stewart, Associate Fire Service Specialist; and Rita Fahy, Ph.D., manager of the NFPA Fire Databases and Systems (and my Irish connection). These folks reviewed manuscript, submitted text and material, and kept the Timex ticking.

Much appreciation to the staff from the Public Fire Protection Division, especially my coworker and technical assistant, Elena Carroll, who kept all the papers and telephone calls shuffled correctly. Thanks for being the lookout.

Since English and editing are not my bailiwick, extra kudos and a tip of the hat to Pam Powell, Product Manager, and Robine Andrau, Senior Developmental Editor, for their devotion to the project and for filling in where the Sisters of St. Joseph tried so many years ago!

Stephen N. Foley

Fire Fighter Fatalities and Injuries in the United States

NFPA has been studying U.S. fire fighter fatalities since 1977, and over that time, there have been significant improvements but also consistent problem areas that still need to be addressed. From 1977 through 2000, almost 2,800 fire fighters in the United States died while on duty.

Over the years, as we've watched the number of fire fighter fatalities decline, we've credited the progress made in reducing fatalities to improvements in fire department apparatus, protective equipment, clothing, training, and incident management. Although we've seen improvements in areas in which solutions could be engineered, we continue to see high proportions of deaths in areas in which changes in behavior are required in order to achieve the same gains. And, until recently, we have not examined the role that the decline in structure fires may have had in explaining the reduction in fire fighter deaths.

Before looking at the problems that still exist, it is important to note the significant improvements in fire fighter health and safety that *have* occurred over the past quarter century. For example, data for the first five years that this study was conducted show an average of 5 deaths a year involving fire fighters who died as a result of exposure to smoke while operating at structure fires. In 1977 alone, 12 fire fighters died as a result of smoke exposure—and 10 of them were not wearing SCBA. From 1996 to 2000, however, we saw only 3 such fatalities.

In a similar vein, with the changes in design of emergency apparatus, fatal falls from apparatus while responding to or returning from emergency incidents have declined dramatically. From 1977 through 1981, the number of fatal falls averaged almost 4 a year (a total of 18 deaths). In contrast, the fatality that occurred in 1999 was the first fatal fall while responding to an emergency since 1991.

Overall, fire fighter fatalities dropped from an average of 151 deaths per year in the late 1970s to 127 deaths per year in the 1980s to 97 deaths per year in the 1990s. That's the good news. But the persistence of two chronic problem areas (heart attacks and motor vehicle-related deaths) and an examination of death rates at structure fires reveal a less positive result.

Heart attacks consistently account annually for more than two-fifths of on-duty fire fighter deaths. Of the almost 2,200 deaths from 1981 through 2000, 45.5 percent

1

were due to heart attacks. Two-thirds of the fire fighters over age 40 who died while on duty died of heart attacks, and the proportion jumps to 3 out of 4 for fire fighters over age 60. Available medical documentation shows that 80 percent of the victims of fatal heart attacks had existing health problems—either prior heart attacks, bypass surgery, or severe, detectable levels of arteriosclerotic heart disease. In spite of their medical conditions, half of the heart attack victims with existing health problems were still actively engaged in fire-fighting activities at the time of their deaths, and another 23 percent were responding to emergencies.

The other chronic problem area concerns motor vehicle crashes, as well as fire fighters struck by vehicles. These deaths account for one-fifth of the deaths over the past 20 years. The motor vehicle crashes most frequently involve volunteer fire fighters responding to emergencies in their own vehicles.

We have recently turned our attention to the trends in fireground fire fighter fatalities. Between 1977 and 2000, the number of deaths at the scene of structure fires dropped by more than one-half. However, over that same time period, the number of structure fires themselves fell by half. As a result, we find that the average death rate in the first five years (1977 through 1981) was 56.2 deaths per million structure fire, and the average death rate from 1996 to 2000 was 52.3 deaths per million structure fires—hardly as dramatic a reduction as we might have expected. Even more alarming is the fact that the highest structure fire death rates occurred in 1989 and 1991 (68.3 and 68.7 deaths per million fires, respectively).

In looking more closely at deaths in structure fires, we find that the death rate for non-heart attack fatalities while fire fighters were operating inside structures has risen from 17.8 deaths per million structure fires to 24.3 deaths per million structure fires. While the causes of fatal injuries to fire fighters who are operating inside structures vary widely, it is possible that changes in equipment and clothing, without compensating improvements in incident management, have exposed fire fighters to greater dangers, with resulting increases in fire fighter death rates.

Chapter 1, "Fire Fighter Fatalities in the United States, 2001," goes into greater detail not only on how many fire fighter fatalities took place in the United States in 2001 but on why these fatalities occurred. Chapter 2, "Fire Fighter Injuries in the United States, 2000," examines the number and types of injuries sustained by fire fighters in the United States in 2000.

Fire Fighter Fatalities in the United States, 2001

Rita F. Fahy and Paul R. LeBlanc

An order comes down, two fire fighters in full-protective clothing and self-contained breathing apparatus head into a burning building to search for survivors. One fire fighter goes for more lights while the other fire fighter continues the search on the third floor. By the time the first fire fighter returns, the fire had spread to the third floor trapping the second fire fighter. As conditions deteriorate, fire crews are ordered out of the building. When the fire is finally extinguished, a rescue team finds their comrade's body in a rear bedroom. Smoke and carbon monoxide asphyxiated him.

As disturbing as this scene is, more disturbing is that during the last several years, fire fighters in the U.S. are dying at structure fires at the same rate as they were in 1977.

Not counting the 340 deaths at the World Trade Center, in 2001 the number of fire fighter deaths at structure fires dropped 58 percent since NFPA began collecting that data in 1977 for its annual report.[1] Improvements in fire-fighting equipment, procedures, and training are commonly cited for what, on the surface, appears to be good news. However, little attention has been paid to a similar decrease of 54 percent in structure fires between 1977 and 2000, the last year for which statistics are available.

Our data indicates that this two-decade decline in structure fires may be what's driving the drop in deaths at such fires (see Figure 1-1). See Figure 1-2 for an overview of on-duty fire fighter deaths.

Rita F. Fahy, Ph.D, is the manager of the NFPA Fire Databases and Systems.

Paul R. LeBlanc is a member of NFPA's Fire Analysis and Research Division and a career lieutenant with the Boston, Massachusetts, Fire Department.

Source: *NFPA Journal*, July/August, 2002.

Credits: This study was made possible by the cooperation and assistance of the United States fire service, the Public Safety Officers' Benefits Program of the Department of Justice, the United States Fire Administration, the Forest Service of the U.S. Department of Agriculture, and the Bureau of Indian Affairs, and the Bureau of Land Management of the U.S. Department of the Interior. The authors would also like to thank Stephen N. Foley and Carl E. Peterson of NFPA's Public Fire Protection Division for their assistance.

[1]NFPA's files for fire fighter on-duty fatal injuries are updated continually for all years.

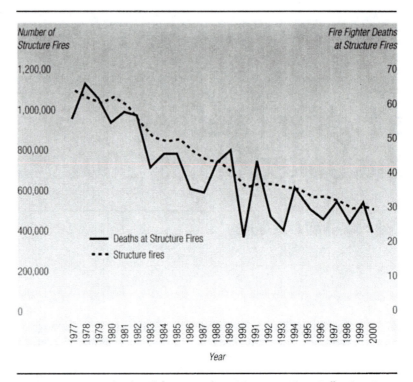

FIGURE 1-1   Drop in Fire Fighter Deaths at Structure Fires Following Drop in Structure Fires

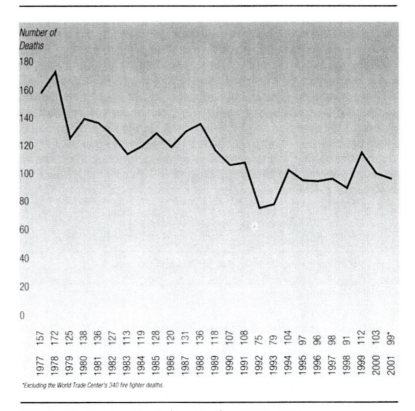

FIGURE 1-2   On-Duty Fire Fighter Deaths, 1977–2001

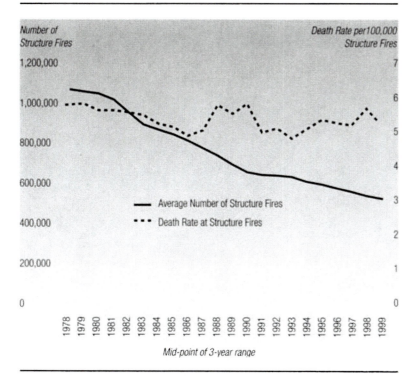

FIGURE 1-3   Comparison of Structure Fires vs. Death Rates of Fire Fighters at Structure Fires (3-Year Running Averages)

We discovered that fire fighter death rates at structure fires declined (see Figure 1-3) until 1987, when they began to rise again. The rate continued increasing, and by the late 1990s, it was roughly the same as it had been in the late 1970s, despite fire-fighting improvements.

We see several possible reasons for this return to a fatality rate of 20 years ago: over reliance on those same advances in protective clothing and equipment; training not improving enough; less experience; and not following accepted incident management protocol.

What injuries are causing these deaths and in what areas might deaths be increasing? Although deaths due to heart attacks declined over the period, the rate of deaths inside structures due to traumatic injuries increased markedly (see Figure 1-4). Sixty-three percent of the deaths that weren't due to heart attacks were the result of smoke inhalation (see Figure 1-5). Burns accounted for another 18 percent, and crushing or internal trauma for 16 percent.

WHY THEY HAPPEN

Most of these traumatic injuries occurred when fire fighters became lost inside a structure, the structure collapsed, or they were overtaken by a rapidly spreading fire, including backdraft and flashover.

Though we saw no consistent trends when looking at cause of injury individually, we noted a clear upward trend when we considered them.

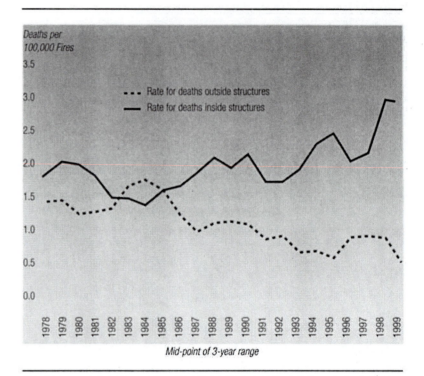

FIGURE 1-4   Rate of Non-Heart Attack Deaths at Structure Fires (3-Year Running Averages)

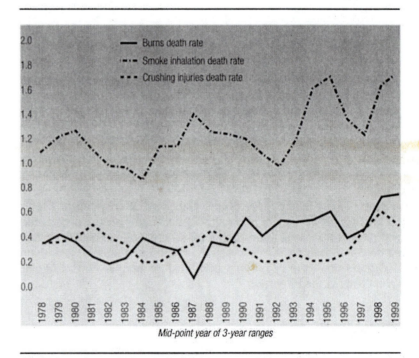

FIGURE 1-5   Rate of Deaths Inside Structure Fires for 3 Major Causes of Death (3-Year Running Averages)

To reduce the number of fire fighter deaths inside structure fires, it's important to understand how they happen and why they're increasing. A detailed look at each incident is beyond the scope of this analysis, but the National Institute for Occupational Safety and Health's (NIOSH) program of investigation of on-duty fire fighter fatalities is providing a database based on on-site data collection. Reports on many of these fatalities can be found at www.cdc.gov/niosh/firehome.html, however, we can provide some general findings.

Of the 87 fire fighters killed since 1990 by smoke inhalation inside structure fires, 29 died when they became lost inside the structure and ran out of air. Another 23 died when they were caught in a rapidly spreading fire fire, a backdraft, or flashover, and 18 died in structural collapses, 10 of them floor collapses. Sixty-nine of these 70 victims were wearing self-contained breathing apparatus. The one exception was a fire fighter who was rescuing his family from a fire in his own home.

Of the 31 fire fighters who died since 1990 of burns they received in structure fires, 14 were caught in, or trapped by, a rapidly spreading fire, backdraft, or flashover, and 12 died in structural collapses.

Finally, 17 fire fighters died since 1990 as a result of crushing injuries or internal trauma, eight in structural collapses.

These statistics raise several important questions. Are fire fighters putting themselves at greater risk while fighting fires inside structures? Do fire fighters think modern protective equipment provides a higher level of protection but don't realize the equipment's limitations or are ignoring those limitations? Have some aspects of modern building contruction or changes in the burning properties of today's contents and furnishings changed the way fires develop?

This area of the fire fighter fatality problem requires closer analysis, but there's still much we can do today to reduce the number of these deaths.

For one thing, we can make sure personnel accountability programs are in place to ensure that incident commanders know where their crews are. And fire fighters must stay with their partners while operating inside structures. If they encounter difficulties, rapid intervention teams (RITs) can be crucial in saving lives, but RITs only work if they can find the endangered fire fighters.

During fire suppression operations, fire fighters must remain highly aware of their surroundings. Conditions can change rapidly, and fire fighters who move too far into a building may find their escape route cut off or too long to traverse. Fire fighters must recognize the danger signs—fires in basements and attics, indications of potential collapses, flashover, and backdraft—and respect them.

They must also heed their low air alarms and turn on their personal alert safety (PASS) devices whenever they enter a structure.

All these recommendations can be found in NFPA's codes and standards. It's crucial to remember, however, that the recommendations work together as a system, and thus rely to a large degree on each other for success. Complying with half the recommendations may not produce half the safety benefit. More than ever, it's clear that fire department safety officers need to guide their departments to full compliance with all safety requirements.

Anecdotally, the fire service is concerned that fire fighters and fire officers may not receive the degree of training and experience necessary to properly assess the risks on the fireground.

If the number of structure fires is decreasing, how do fire fighters and fire officers gain the experience to understand fire progression, fire behavior, and what happens to the structural integrity of a building under fire conditions?

Training provides fire fighters and fire officers an opportunity to learn the intricate and unexact science of fire fighting. Computer and other types of simulations can help and components of the command system, and its risk management decision-making process, can be taught in a classroom simulation environment. A critique of fireground procedures following each major fire is a great opportunity for the crew at the scene and those who weren't at the incident to learn and improve their understanding of fire behavior.

NFPA has standards for training, professional qualifications, and incident management. It's incumbent upon today's fire service leaders to provide the training and the proper promotional assessment processes to ensure that company and chief officers understand the environment their fire fighters are exposed to and the proper operational procedures to deal with that environment so the safety of everyone on the fireground is improved. The fireground is a very unforgiving learning environment.

Incident management, personnel accountability systems, sprinklers and smoke alarms, and the use of PASS devices can reduce fire fighters' risk of dying on duty, but they must also be used routinely to be effective. A comprehensive safety and health program designed using NFPA 1500, *Fire Department Occupational Safety and Health Program*, and its companion standards are essential.

DEATHS BY TYPE OF DUTY

Aside from the deaths at the World Trade Center, the largest proportion of fire fighter deaths in 2001—38 percent—occurred on the fireground, making last year the second consecutive year in which fireground deaths made up less than 40 percent of all on-duty deaths (see Figure 1-6). The largest proportion of fireground deaths occurred in residential structures, in which 16, or 42 percent, of those who died on the fireground perished (see Figure 1-7). Eleven of the 16 died in one- and two-family dwellings, and 5 died in apartment buildings.

We found that, although more fire fighters die in residential structures than in any other type, fires in nonresidential structures, other than educational or health-care facilities and correctional properties, actually pose a greater hazard to fire fighters (see Figure 1-8).

From 1996 to 2000, there were 9.6 fireground deaths per 100,000 nonresidential structure fires compared to 3.7 deaths per 100,000 residential structure fires. The highest death rates occurred in special structures, such as vacant buildings and buildings under construction. The low rate in health-care facilities and in correctional and educational buildings reflects the fact that these occupancies are among the best regulated and most frequently inspected and that their occupants are more likely to report fires in the earliest stages.

The second most common activity resulting in fire fighter deaths in 2001 was responding to or coming from alarms. In the past decade, a fifth to a quarter of all on-duty deaths typically occurred as fire fighters responded to or returned from alarms, and 2001 was no exception. Twenty-four fire fighters died in such a manner last year, 12 due to heart attacks, and one in a crash on the way to an incendiary fire.

Another 23 died while performing non-emergency-related activities. For instance, one fell while doing plumbing work, one was electrocuted as he tried to repair a light fixture at the fire station, and one died when a water tank ruptured.

Twelve fire fighters died during training activities.

Finally, two fire fighters died at non-fire emergencies. One was hit by a pickup truck while directing traffic at the scene of a vehicle crash, and the second died of an aneurysm at another vehicle crash site.

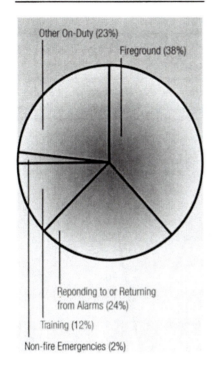

FIGURE 1-6   Fire Fighter Deaths by Type of Duty, 2001

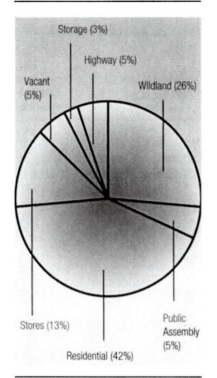

FIGURE 1-7   Fire Fighter Deaths by Fixed Property Use, 2001

Fixed Property Use	Deaths per 100,000 Fires
Vacant, special	18.1

Includes idle buildings, buildings under construction and demolition, etc.

Public Assembly	12.9
Stores/office	12.9
Storage	8.8
Manufacturing	6.0
Residential	3.7
Health-care/correctional	2.4
Educational	0.0

FIGURE 1-8   On-duty Fireground Deaths per 100,000 Structure Fires, 1996–2001

Of the 99 fire fighters who died on duty last year, 87 were members of career and volunteer fire departments, 6 were federal or state contractors, 5 were federal land management agency employees, and 1 was employed by a state forestry agency.[2]

The general trend in career fire fighter deaths since 1985 has been downward, in spite of a rise in deaths from 1993, when fatalities were at their lowest, through 2000. For volunteers, however, the general trend since 1994 has been upward, although the numbers fluctuate a great deal from year to year (see Table 1-1 and Figure 1-9).

FACTORS CONTRIBUTING TO DEATH

For the purposes of our report, we examine the factors that contribute to a death separately from the physical cause of that death. For example, a fire fighter who dies of smoke inhalation may have suffered the smoke inhalation as a result of a structural collapse. In that case, the primary factor leading to his death was the structural collapse, and the cause of his death was asphyxiation.

Let's look at the causal factors first.

---

[2]The term "volunteer" refers here to any fire fighter who isn't a full-time, paid member of a local, municipal fire department. The term "career" refers to full-time, paid, local, municipal fire department members or employees of career organizations whose assigned duties include fire fighting.

TABLE 1-1   Comparison of On-Duty Deaths Between Career and Volunteer Fire Fighters, 2001*

	Career Fire Fighters		Volunteer Fire Fighters	
	Number of Deaths	**Percent of Deaths**	**Number of Deaths**	**Percent of Deaths**
Type of duty				
Operating at fireground	10	41.7	19	30.2
Responding to or returning from alarm	2	8.3	22	34.9
Training	3	12.5	9	14.3
Operating at non-fire emergencies	0	0.0	2	3.2
Other on-duty	9	37.5	11	17.5
Totals	*24*	*100.0*	*63*	*100.0*
Cause of fatal injury				
Stress	10	41.7	30	47.6
Struck by or contact with object	4	16.7	16	25.4
Caught or trapped	8	33.3	11	17.5
Fell	0	0.0	5	7.9
Other	2	8.3	1	1.6
Totals	*24*	*100.0*	*63*	*100.0*
Nature of fatal injury				
Heart attack	10	41.7	30	47.6
Internal trauma	5	20.8	15	23.8
Asphyxiation	5	20.8	6	9.5
Burns	1	4.2	2	3.2
Drowning	0	0.0	3	4.8
Crushing	0	0.0	3	4.8
Gunshot	1	4.2	0	0.0
Other	2	8.3	4	6.3
Totals	*24*	*100.0*	*63*	*100.0*
Rank				
Fire fighter	15	62.5	47	74.6
Company officer	6	25.0	11	17.5
Chief officer	3	12.5	5	7.9
Totals	*24*	*100.0*	*63*	*100.0*
Ages of fire fighters—All deaths				
20 and under	0	0.0	1	1.6
21 to 25	1	4.2	4	6.3
26 to 30	2	8.3	6	9.5
31 to 35	1	4.2	4	6.3
36 to 40	8	33.3	7	11.1
41 to 45	2	8.3	7	11.1
46 to 50	4	16.7	8	12.7
51 to 55	4	16.7	7	11.1
56 to 60	2	8.3	3	4.8
Over 60	0	0.0	16	25.4
Totals	*24*	*100.0*	*63*	*100.0*

TABLE 1-1   (*Continued*)

	Career Fire Fighters		Volunteer Fire Fighters	
	Number of Deaths	**Percent of Deaths**	**Number of Deaths**	**Percent of Deaths**
Ages of fire fighters—Deaths from heart attacks only				
26 to 30	1	10.0	1	3.3
31 to 35	1	10.0	2	6.7
36 to 40	2	20.0	2	6.7
41 to 45	1	10.0	4	13.3
46 to 50	1	10.0	2	6.7
51 to 55	4	40.0	5	16.7
56 to 60	0	0.0	2	6.7
over 60	0	0.0	12	40.0
Totals	*10*	*100.0*	*30*	*100.0*
Fireground deaths by fixed property use				
Dwellings and apartments	5	50.0	11	57.9
Stores	5	50.0	0	0.0
Vacant	0	0.0	2	10.5
Public assembly	0	0.0	2	10.5
Road/highway	0	0.0	2	10.5
Wildland	0	0.0	1	5.3
Storage	0	0.0	1	5.3
Totals	*10*	*100.0*	*19*	*100.0*
Years of service				
5 or less	4	16.7	21	33.3
6 to 10	2	8.3	13	20.6
11 to 15	7	29.2	2	3.2
16 to 20	3	12.5	6	9.5
21 to 25	0	0.0	7	11.1
26 to 30	5	20.8	2	3.2
over 30	3	12.5	10	15.9
Not reported	0	0.0	2	3.2
Totals	*24*	*100.0*	*63*	*100.0*
Attributes of fireground deaths**				
Incendiary and suspicious fires	1		3	
Search and rescue operations	3		0	
Motor vehicle crashes	1		10	
False alarms	0		3	

This table does not include the 12 victims who were contractors for or employees of state or federal wildland agencies. The term volunteer refers to any fire fighter who is not a full-time, permanent, paid member of a fire department.

*This table does not include the 340 New York City fire fighters who were killed in the collapse of the World Trade Center towers on September 11, 2001.

**Since these attributes are not mutually exclusive, totals and percentages are not shown.

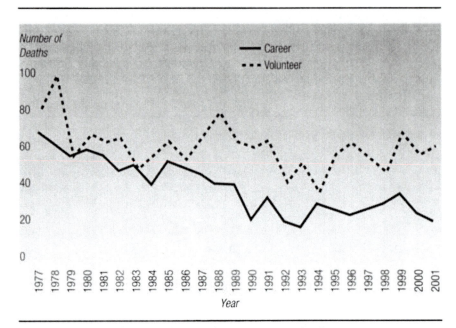

FIGURE 1-9   Fire Fighter Deaths—Local Career vs. Local Volunteer, 1977–2001

As they were in the past, stress and overexertion were the most common factors leading to on-duty deaths, killing 40 fire fighters in 2001 (see Figure 1-10). The second-most common factor was being struck by, or coming into contact with, an object, such as a vehicle or a falling tree. Twenty-seven fire fighters died as a result of this type of trauma. Third on the list was being caught or trapped, which led to 24 fire fighter deaths last year.

Of the remaining 8 fire fighters, 5 died in falls, 1 was electrocuted, 1 had a fatal seizure, and 1 suffered an aneurysm.

CAUSE OF DEATH

The major cause of death was heart attack, even though the number of heart attack deaths has actually dropped by more than one third during the past 25 years. Forty of the on-duty fire fighter deaths in 2001 were the result of heart attacks, and all were attributed to stress or overexertion (see Figure 1-11). Three of the heart attacks occurred as fire fighters responded to false alarms, which were responsible for 27 fire fighter deaths in the last 10 years.

Ten of the 40 victims already had heart problems, usually having suffered heart attacks earlier or having undergone bypass surgery. Eleven had arteriosclerotic heart disease, and three were diabetic. No medical documentation was available for the other 16 victims.

This finding is consistent with the data NFPA has collected since this study began. Over the 25 years we've published this report, medical documentation has been available for 678 of the 1,297 heart attack victims. Of these, 49.4 percent had had previous heart

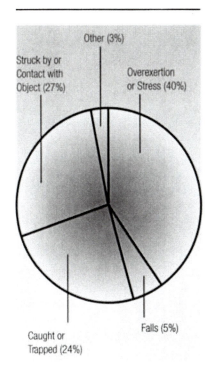

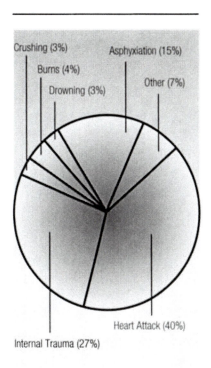

FIGURE 1-10 Fire Fighter Deaths by Cause of Injury, 2001

FIGURE 1-11 Fire Fighter Deaths by Nature of Injury, 2001

attacks or had undergone bypass surgery. Another 30.7 percent had severe arteriosclerotic heart disease, and 12.4 percent had been diagnosed with hypertension or diabetes.

Cardiovascular-related deaths must be reduced. NFPA 1582, *Medical Requirements for Fire Fighters and Information for Fire Department Physicians*, lists the medical conditions that prevent an individual from serving effectively and safely as a fire fighter. Given the large share of deaths among fire fighters with known histories of serious medical conditions, attention to fitness and health throughout a fire fighter's career or volunteer service is essential. The fire service spends a great deal of money on vehicle maintenance. The same care should be devoted to the fire fighters.

Heart attacks tend to hit older fire fighters hardest, killing two out of three of those over 50 in 2001. The youngest heart attack victims were two 27-year-olds, one of whom had no known existing health problems. An autopsy showed that the other had a previously undetected heart defect.

In general, the death rate for fire fighters in their 50s was almost double the all-age average, and for those 60 and older, it was four times the average. Fire fighters older than 50 accounted for two-fifths of fire fighter deaths from 1997 to 2001, although they accounted for less than one-sixth of all fire fighters (see Figure 1-12).[3] As

---

[3]Michael J. Karter, Jr., "U.S. Fire Department Profile Through 1999," NFPA Fire Analysis and Research Division, Quincy, Massachusetts, November 2000, unpublished. The analysis shown here assumes that the number of fire fighters adequately estimates exposure and that the age distribution of career and volunteer fire fighters is similar.

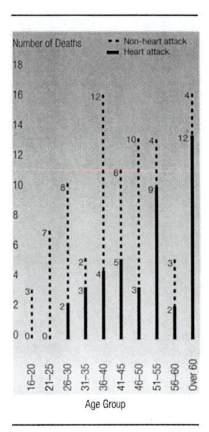

FIGURE 1-12   On-Duty Fire Fighter Deaths by Age and Cause of Death, 2001

might be expected, those in their 30s had the lowest death rate, at about half the average. The fire fighters who died last year ranged in age from 18 to 78, with a median age of 45.

VEHICLE-RELATED INCIDENTS

Seventeen fire fighters died in vehicle crashes in 2001, and vehicles hit and killed four more.[4] Nine of those killed in collisions or rollovers were responding to alarms when the crashes occurred. Of those, five were driving their own vehicles and were the only victim.

Excessive speed was cited as a factor in three of the crashes, and five of the eight fire fighters who died in passenger vehicles or apparatus weren't wearing seatbelts. The driver of one fire engine also tested positive for methamphetamine use.

---

[4]For this report, the term *motor vehicle-related incident* refers to collisions and rollovers involving vehicles, including aircraft and boats, as well as those in which a vehicle played an integral role in the death.

CONCLUSION

The World Trade Center disaster has focused the public's attention on the inherent dangers in fire fighting. It would have been impossible to prevent the catastrophic loss of life that occurred that day, given the scope of the incident. The low death toll of civilians in the buildings below the impact zones is a testimony to the life-saving work of the fire fighters.

Among the 99 other on-duty fatalities that occurred last year, the factors involved in them tended to reflect the same factors we've seen over the past decade. For example, cardiovascular health continues to play a major role in deaths every year, as do crashes while responding to incidents.

Perhaps the most provocative aspect of this year's report is our analysis of structure fire death rates, which shows that the drop in the number of fire fighter deaths during structural fires may, in fact, be the result of the decrease in the number of such fires, rather than improvements in personal protective clothing and equipment, training, and incident management (see Figure 1-13). Since earlier detection should lead to more prompt reporting, sprinkler systems and smoke alarms may have played a role in reducing the incidence or severity of reported structure fires, and fire departments should consider them as part of their public education campaigns.

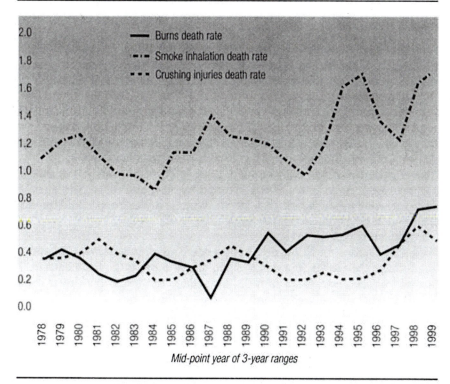

FIGURE 1-13   Rate of Deaths Inside Structure Fires for 3 Major Causes of Death (3-Year Running Averages)

FIRE FIGHTER FATALITIES INCIDENTS 2001

Fire Fighter Killed While Directing Traffic

On January 9, a pickup truck hit a fire fighter directing traffic at the scene of a single-car crash on an icy two-lane highway. He died 10 hours later of his injuries.

The 48-year-old victim and another fire fighter responded to the crash, which involved downed power lines, at 4:42 P.M. To stop traffic from using the road, they parked the apparatus with its emergency lights on in the oncoming travel lane at an intersection.

The two men, neither of whom was wearing protective clothing, were standing next to the apparatus when a car pulled up to the intersection and signaled to turn onto the closed road.

The victim, wearing dark trousers and dark coat, walked over to the automobile and told the driver the road was closed. As he stepped away from the car, he was hit by a pickup traveling in the opposite direction. Road conditions were poor due to snow and ice, and the driver of the pickup was unable to avoid hitting him.

The victim slid 32 feet (10 meters) along the icy road and became trapped under another pickup truck that was stopped in the same lane as the car. Fire fighters and civilians had to lift the truck off the man to get him out.

The fire fighter was taken to the hospital suffering from a massive closed-head injury, a fractured pelvis, pulmonary contusions, and a chest injury. After he was stabilized, he was transferred to a regional trauma center, where he died.

The pickup truck's driver wasn't charged.

Fire Fighter Dies in Tanker Crash

A 21-year-old fire fighter died on January 12 when the tanker he was driving to the scene of a grass fire went out of control. The victim, dressed in bunker gear and a wildland protective coat, was alone in the tanker when the accident occurred.

The fire department was notified of the fire at 2:04 P.M., and the tanker and two engines responded. The tanker was traveling at an undetermined speed along a two-lane road when it drifted to the right as it approached a slight left-hand curve.

The passenger-side tires went off the road, and the tanker traveled in this manner for 100 feet (30 meters) before the driver overcorrected and lost control.

The vehicle crossed the two lanes and drove off the other side of the road into a ditch. It then went up an embankment, hit a utility pole, and rolled over once, ejecting the driver, who wasn't wearing a seatbelt. With the driver trapped underneath, the tanker slid on its side another 30 feet (9 meters).

A fire fighter on his way to the station came across the crash and called 911 at 2:25 P.M. When fire fighters found the driver, they used a hydraulic rescue tool and wood cribbing to raise the tanker so a medic could crawl under it to check on him. More cribbing and airbags were used to raise the vehicle further so the driver could be pulled out. The fire fighter was pronounced dead at the scene.

An autopsy revealed that he'd died of multiple blunt force injuries.

Fire Fighter Dies While Training

A 33-year-old fire fighter died February 5 during physical fitness training.

The victim began his shift at 8 A.M. An hour later, he, a captain, and another fire fighter started their mandatory exercises, which consisted of a half-hour of stretches, followed by a warm-up walk on the treadmill and some weight lifting.

At 10:00 A.M., the captain went to his office, and the other fire fighter went to the kitchen, while the victim did some shoulder muscle resistance exercises. Five minutes later, the fire fighter, who was a paramedic, returned from the kitchen to find the victim lying on the floor.

He noticed that the victim was ashen but he saw no sign of trauma or bleeding. When he couldn't find a pulse, he called to the captain to summon an ambulance.

The ambulance crew spent 15 minutes trying to defibrillate the victim's heart and administering CPR and advanced life support at the station, then took him to the hospital, where he was pronounced dead. An autopsy revealed he'd died of cardiac dysrhythmia associated with exertion due to dilated cardiomyopathy.

Fire Fighter Dies While Hiking

On March 17, a 47-year-old fire fighter collapsed in the middle of a pack test, an arduous three-mile hike that requires the candidate to carry a 45-pound (20-kilogram) backpack the full distance in less than 45 minutes.

Candidates for wildland fire fighting have to pass the pack test to be certified and receive their Red Card.

The victim, who was alone when he collapsed, was discovered by a passerby who called 911. Nearby fire department units responded and took the man to the emergency room, where he was pronounced dead of a heart attack.

The fire fighter had passed the test the previous year.

Fire Fighters Die Attempting Recovery

On April 8, fire fighters were called to a whitewater river in a state park to recover the body of a 23-year-old kayaker, who'd drowned when currents pinned him underwater. First to arrive at 7:30 P.M. were ten members of the department's dive team, including an assistant chief, four divers, two divers in training, and three rope tenders.

After assessing the situation, the assistant chief and a diver entered the water, tethered by a short piece of rope tied to a safety line held by the tenders and several other fire fighters. Both men wore wet suits, dive boots, buoyancy compensators, and masks. Neither was wearing an air cylinder or trim weights.

Although their buoyancy compensators made it difficult to keep their heads above water, both managed to reach the kayaker. Shortly afterward, however, the fire fighter's safety line became tangled between a tree and the kayaker's body. The assistant chief tried to free it, but couldn't, and after two rescue attempts, both men went under the water.

When efforts to pull them ashore failed, the crew on the bank cut the safety line, and both men reappeared, floating downstream, and still tethered together. Their bodies were recovered 500 yards (450 meters) from the site of the failed recovery attempt. Neither man had a pulse when pulled from the water.

Fire fighters started CPR and defibrillated the victims several times before carrying them up a 200-foot (60-meter) embankment to advanced life support ambulances. Both men were pronounced dead at the hospital.

Fire Fighter Dies After Fall

At 9:15 A.M. on April 10, a fire fighter trying to fix a light fixture 15 feet (5 meters) above the fire station floor died when he fell from a ladder after being electrocuted.

The victim, a former electrician, had decided to fix the fixture when a new bulb failed to do the trick. With the help of a colleague, he placed a 24-foot (7-meter) extension ladder, partially extended, against a rafter and climbed up to the fixture. Standing with his back to the rungs, he started work on the light while the other man held the base of the ladder.

When the man on the ground saw the victim slump forward, he tried to break his fall, but the victim hit the concrete floor head and shoulder first. Hearing the ladder fall, a third fire fighter ran into the main floor and saw the victim lying on the floor, bleeding from the head. He called the station ambulance crew, and they began trying to resuscitate the victim, whose heart had stopped. The man was pronounced dead at the hospital.

Tree Kills Fire Fighter

On July 10, four fire fighters were performing routine fire hydrant maintenance when a 3,760-pound (1,700-kilogram) tree fell on one of their vehicles, killing one fire fighter and injuring another.

The fire fighters had split into two crews to perform their task. One crew was given a staff car and told to inspect each hydrant visually, making sure the caps were free. They were then to open one cap and flush the hydrant. The second crew, using a box-style ambulance without a walk-through, was to wait until clear water was flowing, then shut the hydrant down, lubricate the caps, and move on to the next hydrant.

The second crew had just closed a hydrant and was driving slowly down the street when, for no apparent reason, a dead tree fell on the ambulance. It hit the vehicle diagonally, from the rear of the cab's passenger side to the front on the driver's side, and flattened the cab to the dashboard.

The fire fighter in the passenger seat was crushed. The driver, though hit by the roof, was partially ejected from the ambulance when his door popped open, giving him just enough space to survive. Although he was dazed and disoriented, he managed to undo his seatbelt, get out of the vehicle, and call for help.

The first crew, both paramedics, arrived within seconds and treated the driver, who'd suffered a concussion. However, they were unable to get to the victim, who was trapped in the cab under a 2,770-pound (1,256-kilogram) piece of the tree. A wrecker had to be brought in to lift the tree off the vehicle, and even then, rescuers could only reach the victim's left hand and wrist.

She had no pulse and was pronounced dead at the scene. An autopsy revealed that she'd died instantly as a result of compressional asphyxia.

Tree Falls on Fire Fighter

Lightning from a passing thunderstorm hit a fir tree in a wildland area on August 31, and burning embers fell to the ground, where they smoldered until the surrounding trees ignited. The fire was spotted September 3, when a passerby saw it and called 911 at 1:20 A.M.

A 20-person fire crew arrived on the scene at 1:40 A.M. to find flames enveloping the top third of a tree-covered slope. After sizing up the fire, the incident commander decided the crew should stay on the fireground and monitor the blaze until dawn, when they could start extinguishing it.

In the morning, fire fighters began cutting down trees they thought might fall and injure the crews cutting firebreaks. At 9 A.M., one of them put his saw down to get a drink of water, when a tree 40 feet (12 meters) tall and 8 inches (20 centimeters) in diameter fell on him. He died of head and chest injuries at the hospital.

Captain Kills Assistant Chief, Then Self

On September 13, an assistant fire chief went to the house of a troubled fire captain to accompany him to fire department headquarters for a 9:30 A.M. meeting to discuss his problems.

In front of his wife, the captain shot the assistant chief twice in the chest, then turned the gun on himself. The men, long-time friends, died at the house.

Fire Fighter Dies in Blaze

At 2:25 P.M. on November 10, fire crews were dispatched to a 40-acre (16-hectare) wildland fire they believed had been contained.

A four-person crew was sent to the flank of the fire to cut a firebreak by hand because the terrain was so steep it wasn't safe to use a bulldozer. The bulldozer was driven to the other flank to clean out an old logging road and create a back-up firebreak.

Using two leaf blowers and two fire rakes, the crew worked down the slope along a drain, when the fire fighter farthest down saw a glow below him in the drainage ditch. Shortly afterward, he saw a fire blowing up the drain toward him and told the fire fighter next to him that they had to get out. He then ran up the slope to warn the other two crew members, stopping to take off his leaf blower.

He made it to safety, with minor facial burns, as did two other fire fighters. However, the flames overran the fourth man, who was pronounced dead at the scene of smoke inhalation and burns.

The 46-year-old victim, a three-year seasonal employee, was dressed in protective clothing, but he had no goggles or fire shelter. None of the other crew members had shelters, either, although they were available.

The size and slope of the drainage ditch, the abnormally dry fuel load, and the large rocks that made moving about the ditch difficult all contributed to his death.

Fire Fighter Succumbs to Injuries

On January 1, a passerby called 911 at 2:15 A.M. to report a fire in a single-family house. The one-and-a-half story, wood-frame dwelling covered 1,600 square feet (1,338 square meters).

The fire chief, who was first on the scene, noted a lot of smoke, an orange glow behind the basement windows, and fire on the first floor. Hearing that residents might still be inside, he ordered the first-arriving company to conduct primary search.

Two fire fighters from the first engine company, wearing full protective ensembles, and positive-pressure self-contained breathing apparatus, entered the rear of the house. As they did, the floor collapsed, dropping one man into the basement. The second managed to keep from falling, but he was exposed to flames and superheated air rising from the basement.

As soon as they realized what had happened, other fire fighters began a rescue attempt, but the furniture and appliances that had fallen onto the man in the basement when the floor collapsed hampered their efforts. When he was finally freed, the victim was taken to the hospital, then transferred to a burn center. The second fire fighter was also taken to the hospital, where he was treated for minor burns to his face and right arm, and released.

The victim, who sustained second- and third-degree burns over 75 percent of his body, underwent numerous operations, before he died on March 25.

Investigators determined that the fire started in the basement near the clothes dryer. The residents said that there'd been a fire in the dryer the evening before, but they thought they'd extinguished it.

Trapped Fire Fighter Dies

At 8:37 P.M. on May 9, a man called 911 from a neighbor's house to report a fire in his apartment building. The three-story structure was of ordinary construction, contained 13 apartments, and had a ground-floor area of 2,250 square feet (209 square meters). Its common area was equipped with hardwired smoke detectors, and each apartment had battery-operated smoke detectors, which operated.

Three responding engines and a ladder truck reported fire showing from the second story. When residents informed the fire chief that three other residents were trapped on the second floor, he ordered the ladder company to search for them.

Despite the fire's growing intensity, two fire fighters in full-protective clothing and self-contained breathing apparatus went to the second floor to begin the search. When they failed to find anyone, they returned to the second-floor hallway, and one went to get more lights. The second fire fighter went to the third floor to search.

By the time the first man returned, the fire had spread to the third floor and the cockloft, trapping the second fire fighter.

Intense heat and flames stymied several rescue attempts, and the trapped man soon radioed that his air supply was running out. As conditions deteriorated, fire crews were ordered out of the building, and the aerial ladder was used as a water tower. When the fire was finally extinguished, a rescue team found the victim's body in a bedroom at the rear of the building. Smoke and carbon monoxide had asphyxiated him.

Investigators determined that the fire started in a bedroom in a second-floor apartment, reportedly when an unattended child dropped a flaming piece of paper onto a pile of clothing after it burned her fingers. She'd lit the paper with a candle so she could light the gas stove in the kitchen.

Fire Fighter Injuries in the United States, 2000

Michael J. Karter, Jr. and Stephen G. Badger

In 2000, fire fighter injuries in the United States decreased 4.4 percent from the year before, to 84,550 injuries (see Figure 2-1). This is the fewest reported since 1977, when 106,100 nonfatal injuries were recorded. At the time, exposure to infectious diseases such as hepatitis, meningitis, and HIV was still included in the general injury count. However, such exposure was given its own category in 1993. By 2000, NFPA estimates that there were 11,500 exposures to infectious diseases, or 0.9 exposures per 1,000 emergency medical runs. This is down from 12,700 fire fighter exposures in 1999.

Also down from 1999, by 21.1 percent, was the number of fire fighters who had to be hospitalized as a result of their injuries. An estimated 3,650 of the fire fighters injured last year were hurt badly enough to be hospitalized.

Fifty-one percent of these injuries, or 43,065, occurred during fireground operations. Another 15,725, or 18.6 percent, occurred during other on-duty activities, while 13,660, or 16.2 percent, occurred at nonfire emergencies (see Figure 2-2).

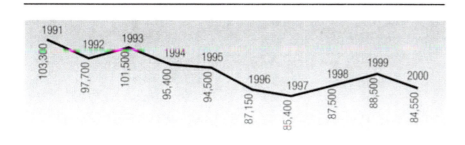

FIGURE 2-1 Total Fire Fighter Injuries by Year (1991–2000)
Source: *NFPA Annual Survey of Fire Departments for U.S. Fire Experience (1991–2000)*

Michael J. Karter, Jr. is a senior statistician with NFPA's Fire Analysis and Research Division. Stephen G. Badger is a member of NFPA's Fire Analysis and Research Division and a full-time fire fighter with the Quincy, Massachusetts, Fire Department.

Source: *NFPA Journal*, November/December, 2001.

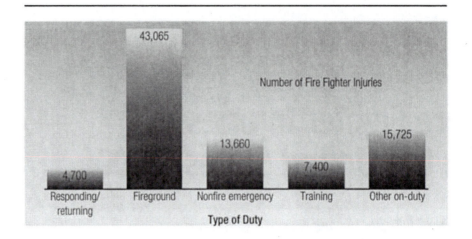

FIGURE 2-2   Fire Fighter Injuries by Type of Duty, 2000
Source: *NFPA Annual Survey of Fire Departments for U.S. Fire Experience (1991–2000)*

Between 1988 and 1997, fire fighter injuries at the fireground decreased 33.8 percent, from 61,790 to 40,920 (see Table 2-1). However, the rate of injuries per 1,000 incidents remained fairly constant, except in 1996 and 1997, because the incidents also decreased, by 29.9 percent.

TABLE 2-1   Fire Fighter Injuries at the Fireground and at Nonfire Emergencies, 1988–2000

Year	Injuries at the Fireground	Injuries per 1,000 Fires at the Fireground	Injuries at Nonfire Emergencies	Injuries per 1,000 Incidents at Nonfire Emergencies
1988	61,790	25.4	12,325	1.13
1989	58,250	27.5	12,580	1.11
1990	57,100	28.3	14,200	1.28
1991	55,830	27.3	15,065	1.20
1992	52,290	26.6	18,140	1.43
1993	52,885	27.1	16,675	1.25
1994	52,875	25.7	11,810	0.84
1995	50,640	25.8	13,500	0.94
1996	45,725	23.1	12,630	0.81
1997	40,920	22.8	14,880	0.92
1998	43,080	24.5	13,960	0.82
1999	45,500	25.0	13,565	0.76
2000	43,065	25.2	13,660	0.73

Source: *NFPA Annual Survey of Fire Departments for U.S. Fire Experience (1988–2000)*

As is generally the case, the four major types of fireground injuries in 2000 were strains and sprains, wounds, burns, and smoke or gas inhalation. Strains and sprains comprised 45.3 percent of the injuries, while wounds, cuts, bleeding, and bruises comprised 17.5 percent. Burns accounted for 8.9 percent, and smoke or gas inhalation comprised 6.7 percent. As for nonfireground injuries, strains, sprains, and muscular pain accounted for 56.1 percent, while wounds, cuts, bleeding, and bruises accounted for 16.9 percent (see Table 2-2).

The leading causes of the fireground injuries were overexertion and strain, which were responsible for 31.4 percent, and falls, slips, and jumps, which accounted for 25 percent (see Figure 2-3). Another 14.2 percent of the fireground injuries were due to exposure to fire products, while 8.9 percent resulted when fire fighters stepped on, or came in contact with, an object (see Figure 2-4).

About 1.2 percent of all fire fighter injuries, or 15,300 injuries, were the result of collisions involving fire department emergency vehicles (see Table 2-3). Because fire departments responded to more than 20.5 million incidents in 2000, the number of collisions represents only about one-tenth of 1 percent of total responses. In addition, 1,160 collisions involving fire fighters' personal vehicles resulted in 170 injuries.

TABLE 2-2 Fire Fighter Injuries by Nature of Injury and Type of Duty, 2000

Nature of Injury	Responding to or Returning from an Incident		Fireground		Nonfire Emergency		Training		Other On-Duty		Total	
	Number	%	Number	%	Number	%	Number	%	Number	%	Number	%
Burns (Fire or Chemical)	75	1.6	3,850	8.9	100	0.7	280	3.8	195	1.3	4,500	5.3
Smoke or Gas Inhalation	145	3.1	2,870	6.7	275	2.0	80	1.1	65	0.4	3.435	4.1
Other Respiratory Distress	95	2.0	855	2.0	180	1.3	70	0.9	270	1.7	1,470	1.7
Eye Irritation	135	2.9	2,220	5.2	285	2.1	220	3.0	520	3.3	3,380	4.0
Wound, Cut, Bleeding Bruise	875	18.6	7,525	17.5	1,840	13.5	1,270	17.2	3,025	19.2	14,535	17.2
Dislocation, Fracture	185	3.9	1,170	2.7	215	1.6	295	4.0	525	3.3	2,390	2.8
Heart Attack or Stroke	75	1.6	250	0.6	135	1.0	110	1.5	405	2.6	975	1.1
Strain, Sprain Muscular Pain	2,500	53.2	19,500	45.3	8,125	59.5	4,350	58.8	8,285	52.7	42,760	50.6
Thermal Stress (frostbite, heat exhaustion)	85	1.8	2,175	5.0	100	0.7	285	3.8	115	0.7	2,760	3.3
Other	530	11.3	2,650	6.1	2,405	17.6	440	5.9	2,320	14.8	8,345	9.9
	4,700		43,065		13,660		7,400		15,725		84,550	

Source: *NFPA's Survey of Fire Departments for U.S. Fire Experience (2000)*. Note: If a fire fighter sustained multiple injuries for the same incident, only the nature of the single most serious injury was tabulated.

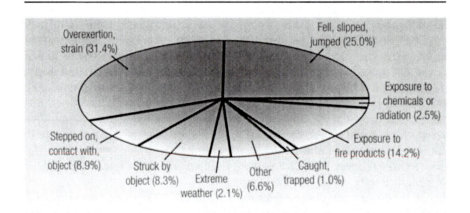

FIGURE 2-3   Fireground Injuries by Cause, 2000
Source: *NFPA's Annual Survey of Fire Departments for U.S. Fire Experience (2000)*

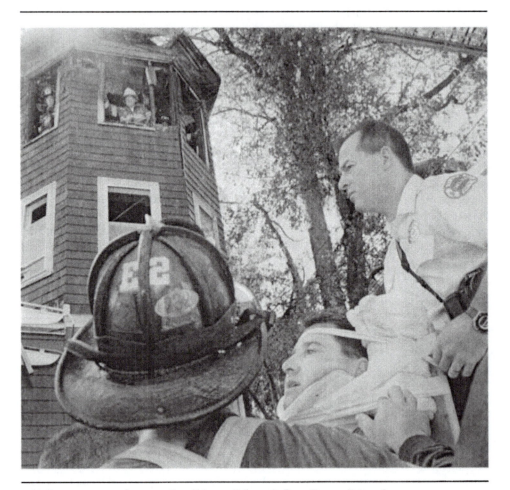

FIGURE 2-4   Massachusetts Fire Fighter Hit by a Falling Piece of Wood
Source: Copyright © AP/Wide World

TABLE 2-3   Fire Department Vehicle Accidents and Resulting Fire Fighter Injuries While Responding to or Returning from Incidents, 1990–2000

Year	Collisions Involving Fire Department Emergency Vehicles	Fire Fighter Injuries Involving Fire Department Emergency Vehicles	Collisions Involving Fire Fighters' Own Vehicles	Fire Fighter Injuries Involving Fire Fighters' Own Vehicles
1990	11,325	1,300	950	175
1991	12,125	1,075	1,375	125
1992	11,500	1,050	1,575	150
1993	12,250	900	1,675	200
1994	13,755	1,035	1,610	285
1995	14,670	950	1,690	190
1996	14,200	910	1,400	240
1997	14,950	1,350	1,300	180
1998	14,650	1,050	1,350	315
1999	15,450	875	1,080	90
2000	15,300	990	1,160	170

Source: *NFPA's Survey of Fire Departments for U.S. Fire Experience (1990–2000)*

INJURIES BY POPULATION AND REGION

Table 2-4 breaks down the average number of fires and fireground injuries per department by region of the country and by the size of the population of the community protected. The four regions of the United States, as defined by the U.S. Bureau of the Census, are

- *Northeast:* Connecticut, Maine, Massachusetts, New Hampshire, New Jersey, New York, Pennsylvania, Rhode Island, and Vermont.
- *North Central:* Illinois, Indiana, Iowa, Kansas, Michigan, Minnesota, Missouri, Nebraska, North Dakota, Ohio, South Dakota, and Wisconsin.
- *South:* Alabama, Arkansas, Delaware, District of Columbia, Florida, Georgia, Kentucky, Louisiana, Maryland, Mississippi, North Carolina, Oklahoma, South Carolina, Tennessee, Texas, Virginia, and West Virginia.
- *West:* Alaska, Arizona, California, Colorado, Hawaii, Idaho, Montana, Nevada, New Mexico, Oregon, Utah, Washington, and Wyoming.

Where fire departments reported sufficient data by region, the Northeast had substantially more fireground injuries for communities of almost every size than elsewhere in the United States. At 5.7 injuries per 100 fires, the Northeast reported more than twice the rate of fireground injuries as the rest of the country.

TABLE 2-4   Average Number of Fires and Fireground Injuries per Department and Injuries per 100 Fires by Population of Community Protected and Region, 2000

Population of Community Protected	Northeast			North Central			South			West		
	1	2	3	1	2	3	1	2	3	1	2	3
500,000 to 999,999	*	*	*	*	*	*	2,786.0	114.3	3.0	2,400.8	65.5	2.7
250,000 to 499,999	*	*	*	2,497.8	76.0	3.0	1,868.6	32.5	1.7	1,800.3	31.9	1.8
100,000 to 249,999	*	*	*	684.2	18.9	2.8	982.6	16.6	1.7	507.8	13.5	2.7
50,000 to 99,999	381.8	22.5	5.9	273.1	6.8	2.5	386.3	6.8	1.8	320.1	5.5	1.7
25,999 to 49,999	145.3	4.7	3.3	128.4	2.9	2.3	179.8	2.4	1.3	189.8	2.7	1.4
10,000 to 24,999	59.3	2.8	4.7	71.8	1.5	2.1	111.1	1.3	1.2	102.7	1.0	1.0
5,000 to 9,999	35.2	1.5	4.3	41.7	0.7	1.7	58.5	1.2	2.1	51.8	1.3	2.5
2,500 to 4,999	21.9	0.8	3.7	25.8	0.5	1.9	35.7	0.5	1.4	24.8	0.6	2.4
Under 2,500	9.0	0.3	3.3	13.2	0.2	1.5	21.0	0.3	1.4	10.9	0.3	2.8
Overall Regional Rate			5.7			2.0			1.6			2.1

Column 1: Average Reported Number of Fires
Column 2: Average Reported Number of Fireground Injuries
Column 3: Number of Fireground Injuries per 100 Fires

Source: *NFPA's Survey of Fire Departments for U.S. Fire Experience (2000)*

*Insufficient data

      Table 2-5 indicates that the number of fires to which a fire department responds is directly related to the size of the population it protects and that the number of fires a department attends has a direct bearing on the number of fireground injuries its personnel incur. The second point is clearly demonstrated by the range of injury numbers, which run from a high of 112.1 for departments protecting communities of 500,000 to 999,999 to a low of 0.2 for departments that protect communities of less than 2,500 (see Table 2-5).

TABLE 2-5   Average Number of Fires, Fireground Injuries, and Injury Rates by Population of Community Protected, 2000

Population of Community Protected	Average Number of Fires	Average Number of Fireground Injuries	Number of Fireground Injuries per 100 Fires	Number of Fireground Injuries per 100 Fire Fighters
500,000 to 999,999	3,576.9	112.1	3.1	17.3
250,000 to 499,999	1,977.5	48.7	2.5	9.2
100,000 to 249,999	808.5	19.0	2.4	7.6
50,000 to 99,999	327.1	8.0	2.4	7.6
25,000 to 49,999	153.4	2.9	1.9	6.2
10,000 to 24,999	82.4	1.6	1.9	3.8
5,000 to 9,999	45.4	1.0	2.2	2.9
2,500 to 4,999	27.4	0.5	1.8	2.4
Under 2,500	13.6	0.2	1.5	1.3

Source: *NFPA's Survey of Fire Departments for U.S. Fire Experience (2000)*

Community size also influences fire fighter hospitalizations. The number of injuries for which fire fighters in departments protecting communities of 250,000 or more had to be hospitalized in 2000 ranged from a high of 6.69 for departments protecting communities of 500,000 to 999,999 to a low of 2.39 for those protecting communities with populations of 250,000 to 499,999. For departments protecting communities of less than 250,000 people, the average number of injuries requiring hospitalization was less than 1 for all population groupings.

Relating the number of injuries requiring hospitalization to the total number of injuries that occur in a department allows more direct comparison with the severity of injuries suffered by fire fighters in departments protecting communities of different sizes. The number of injuries requiring hospitalization per 100 injuries for departments by population of community is shown in Table 2-4. Overall, departments that protect the smallest communities—those with populations under 25,000—incurred 7.5 injuries requiring hospitalization per 100 injuries sustained. This is more than twice the rate of departments that protect communities with populations of 25,000 to 999,999.

Table 2-6 also indicates the average number of injuries incurred in fire departments in communities of different sizes. Overall, the larger the community, the higher the number of injuries The average number of injuries per department ranged from a high of 334.74 to a low of 0.32 for departments that protect communities with populations under 2,500.

The average number of fires and fireground injuries per department by the size of the population of the community protected in 2000 is shown in Table 2-5, which reveals the number of fireground injuries per 100 fires.

Examining the number of fireground injuries that occur for every 100 fires attended is useful in determining fire fighter injury experience and obtaining a reading on the risk departments face. This allows one to take into account relative fire experience

TABLE 2-6   Average Number of Fire Fighter Injuries and Injuries Requiring Hospitalization per Department, by Population of Community Protected, 2000 (for all types of duty)

Population of Community Protected	Average Number of Fire Fighter Injuries	Average Number of Injuries Requiring Hospitalization	Number of Injuries Requiring Hospitalization per 100 Injuries
500,000 to 999,999	334.74	6.69	2.00
250,000 to 499,999	122.90	2.39	1.94
100,000 to 249,999	58.52	0.89	1.53
50,000 to 99,999	21.19	0.56	2.64
25,000 to 49,999	6.19	0.20	3.23
10,000 to 24,999	2.93	0.18	6.14
5,000 to 9,999	1.52	0.11	7.24
2,500 to 4,999	0.78	0.07	8.97
Under 2,500	0.32	0.04	12.50

Source: *NFPA's Survey of Fire Departments for U.S. Fire Experience (2000)*

and provides a direct comparison of departments protecting communities of different sizes.

The overall range of rates of fireground injuries per 100 fires varied little from a high of 3.1 for departments protecting communities of 500,000 to 999,999 to a low of 1.5 for those protecting communities of less than 2,500. Thus, the wide range noted in average fireground injuries by population protected narrows when relative fire experience is taken into account. The overall injury rate of departments protecting communities of 50,000 or more was 3.7 injuries per 100 fires, or 60 percent higher than the rate of departments protecting communities of less than 50,000.

The risk of fireground injury per 100 fire fighters by size of community protected was also calculated. Larger departments generally had the highest rates, with departments protecting communities of 500,000 to 999,999 having the highest rate at 17.3 injuries per 100 fire fighters. As community size decreases, the rate drops to 1.3 for departments protecting fewer than 2,500 people. That's a 13-to-1 difference in risk of injury between communities of 500,000 to 999,999 and those with less than 2,500 inhabitants.

Although a department protecting a community with a population of 500,000 to 999,999 has, on average, more than 50 times as many fire fighters than does a department protecting a population of less than 2,500, the larger departments attend more than 260 times as many fires. As a result, they incur considerably more fireground injuries.

Improving Fire Fighter Safety

It's unlikely that all fire fighter injuries can be eliminated. However, a risk management system and the application of existing technology can reduce injury levels, lost time, and medical costs. A safety and health program is also beneficial in reducing injuries. NFPA 1500, *Standard on Fire Department Occupational Safety and Health Program*, is a good place to begin when developing such safety and health programs.

At the local level, departments can make a commitment to reducing injuries, establish a safety committee, use appropriate protective equipment, develop a program on using and maintaining self-contained breathing apparatus (SCBA), enforce safe driving policies for operators and passengers of fire apparatus, ensure sufficient personnel for fire-fighting and overhaul duties, and implement medical examinations and a physical fitness program.

Departments should also consider introducing an incident management system, if they haven't already done so, and establishing programs to help communities install private fire protection systems. Such fire protection systems will allow fires to be discovered and reported earlier, reducing fire fighters' exposure to a hostile environment.

INCIDENTS

Texas: Trapped Fire Captain and Rescuer Burned

Fire fighters responded to a 911 call of a fire in a two-story apartment of unprotected wood-frame construction at 4:26 A.M.

Arriving fire crews advanced a hose line to the second floor, where they encountered heavy smoke, but the hose line was too short to reach the fire. While they were trying to extend the hose line, a fire captain wearing a full-protective ensemble climbed to the second story, where he donned his SCBA face piece and entered the apartment to perform a primary search.

As conditions deteriorated, he became trapped. Hearing the captain's scream for help, a fire fighter looked up and saw flames coming from the apartment window. The fire fighter grabbed an unused hose line and knocked down the fire, then climbed a ladder and helped the captain to a window, where another fire fighter waited to take him down. At that point, the captain tumbled down the ladder, but a fire fighter broke his fall.

The captain was hospitalized for two months with second- and third-degree burns to 53 percent of his body. He missed 18 months of duty but returned in a staff support position. Investigators believe his protective ensemble saved his life.

The rescuer, who jumped from the window, suffered second- and third-degree burns to his forearms, neck, and ears. He was hospitalized for two days and missed 25 days of duty. Investigators believe his protective ensemble prevented more serious injuries.

A company officer who suffered minor smoke inhalation was hospitalized for a day and missed four days of duty.

The cause of the fire is undetermined.

California: Fire Fighter Disoriented During Flashover

Fire fighters responding to a 911 call at 10:30 P.M. found a fire burning in the basement of a two-story, single-family dwelling of unprotected ordinary construction and smoke filling the house.

A fire fighter in full-protective ensemble and SCBA was searching the structure for a victim when smoke and gasses ignited and flashed over, disorienting and trapping him. He was treated at the hospital for a first-degree burn on his arm and released the same day. He was later admitted for another three days as the result of an infection.

The injured man missed 10 tours of duty, then worked 8 days of light duty before returning to his regular assignment. Fire officials credit his ensemble with keeping his injuries to a minimum.

Another fire fighter was dehydrated, and a company officer sprained an ankle. Neither lost any time from work.

The fire, which began when an open flame in the basement ignited natural gas vapors escaping from a furnace, spread to the house's structural members.

Pennsylvania: Fire Chief Shocked Battling Chimney Fire

In mid-December, fire fighters responded to a 911 call reporting a chimney fire in a two-story, single-family house of unprotected wood-frame construction.

Wearing a full-protective ensemble and safety glasses but no SCBA, the fire chief climbed an extension ladder to extinguish the fire. He was using brush and steel rods to clean the chimney when the rods touched a 7,800-volt power transmission line. The jolt of current knocked the chief off balance, and he fell 14 to 16 feet (4 to 5 meters) backwards, hitting his head on a lower rung of the ladder.

The impact fractured his cervical spine, injured his central nervous system, and severely lacerated his scalp. He was hospitalized for 56 days and evaluated for electric shock, burns, and spinal cord injuries. He needed 42 stitches to close his head laceration and was diagnosed as a quadriplegic. He is currently undergoing therapy three days a week and recovering gradually.

Creosote in the chimney caused the fire.

Delaware: Assistant Fire Chief Injured Evacuating

At 7:09 on an evening in late October, fire fighters responding to a chimney fire in a two-story townhouse found smoke showing and fire extending to the roof.

The assistant chief, wearing full protective ensemble and an SCBA, went into the second story with a hose line. He knocked down the fire in a bedroom, then attacked the fire on the exterior of the house by leaning out a window. As he did so, he noticed fire coming from the window below him.

At about the same time, a fire fighter in the room with the assistant chief opened the room's ceiling to check for fire extension and was met with heavy black smoke and high heat. The assistant chief then put down his nozzle and went to the door.

At that point, deteriorating conditions forced the incident commander to order an evacuation alarm. The assistant chief heard the order and the sirens, and he called to the two fire fighters with him. He forced himself through the intense heat down the stairs to the bottom, where the cooler air led him to believe he'd made it outside. Only after he removed his helmet, hood, and face piece and took a breath did he realize he was still inside the house. He immediately broke a window with his helmet and tried to climb out as others ran to help him.

The assistant chief was hospitalized for 19 days and placed on a ventilator. He missed 60 days of duty, then returned to light duty for 60 tours. He has since returned to full duty.

Two other fire fighters were treated for first- and second-degree burns.

A faulty chimney caused the fire.

New York: Blaze Traps Fire Fighter

At 12:20 in the afternoon in early March, fire fighters responded to a telephone call reporting a fire on the fifth story of a six-story apartment building with 36 units. The structure was of protected, ordinary construction.

Fire fighters found smoke in an apartment on the fourth story and discovered that a fire burning in a bucket of floor sealer in the kitchen had spread to the kitchen cabinets. Evidently, the apartment floors were being refinished, and workers had left the lower halves of the windows open.

A fire fighter in a full-protective ensemble and SCBA was ordered to extinguish the blaze, while other crew members searched the premises. However, conditions deteriorated immediately, and the fire fighter left the apartment as the fire began to spread rapidly.

The fire fighter suffered third-degree burns to the back of the head and shoulders, and second-degree burns to his left hand and forehead. He was admitted to the hospital, where he stayed for 52 days, 35 of which he was in a coma. Investigators believe he was wearing his protective hood under his SCBA face piece instead of over it and that his chinstrap wasn't on properly, allowing his helmet to fall off. He missed 511 days of duty but has since returned to full duty.

The fire began when heat from some type of electrical equipment ignited flammable floor sealer vapors.

Wisconsin: Circular Saw Cuts Fire Fighter's Leg

Fire fighters responded to a mini-bus fire at 4:48 P.M. in March and found a vehicle fire had spread to a single-family house of unprotected wood-frame construction.

To save the house, a fire fighter in full protective ensemble and wearing, but not using, SCBA climbed onto its pitched roof with a gas-powered circular saw to remove smoldering roofing materials from the eaves. As he worked from a roof ladder, the saw jumped and fell on his outstretched leg.

The fire fighter was hospitalized for an unreported number of days for a deep laceration to the hamstring and muscle tissue. Though he underwent extensive therapy, he was unable to return to duty and retired 11 months later on a medical disability.

The fire was incendiary.

California: Fire Fighter Suffers Smoke Inhalation

At 7:40 on a January morning, fire fighters responded to a 911 call for a fire in a two-story, four-family townhouse of protected ordinary construction.

A fire fighter in a full-protective ensemble and SCBA was helping lay out a hose line for an interior fire attack when his low-air alarm sounded. He got help replacing his air cylinder, then resumed work, pairing with another fire fighter to perform a second-story search.

During the search, furnishings fell on his back, and he inadvertently disconnected his hose line while trying to remove them. The fire fighter soon became disoriented, and his low-air alarm began sounding again. His air supply depleted, he removed his face piece and yelled for help. His partner and a rapid intervention team heard his cries and removed him, by now unresponsive, from the house.

The fire fighter was hospitalized for four days, suffering from smoke inhalation. He missed 34 days of work, then returned to regular duty.

Investigators found a small rock in the exhaust valve of his SCBA, which accelerated the air leakage. It wasn't reported how the rock got there.

Another fire fighter who responded to the house fire, also in a full-protective ensemble and SCBA, was burned on the neck by hot roofing tar when he removed his helmet to free himself after becoming tangled in ceiling material. He missed no duty.

Combustibles too close to a heater in a closet started the fire.

Fire Department Occupational Safety and Health Programs

There are many components to an overall fire department's occupational safety and health program. These components are created by individuals and by committees, all of whom have a responsibility to ensure the health and safety of fire department members. The roles and responsibilities of these individuals and committees are framed by the requirements in NFPA 1500, *Standard on Fire Department Occupational Safety and Health Program*, but they are enforced by individual organizations and/or the authority having jurisdiction.

Chapter 3, "Fire Service Occupational Safety, Medical, and Health Issues" begins with an overview and history of the involvement of NFPA and other regulatory agencies in fire department health and safety programs.

Chapter 4, "Evaluation and Planning of Public Fire Protection," spells out the various roles and responsibilities. These roles and responsibilities must be outlined by the organizations or individuals involved, that is, the fire chief, the health and safety officer, the fire department occupational safety and health committee, and the various represented organizations or associations.

Chapter 4 also presents a framework for how to evaluate the services the department provides and how the risk-versus-benefit model is employed internally and externally. This risk-versus-benefit model is important because it impacts the health and safety of fire fighters both administratively and operationally. Fire departments are required in NFPA 1500, *Standard on Fire Department Occupational Safety and Health Program*, to operate with a risk management matrix that defines what risks the fire department will respond to within a community. This risk management plan includes input from citizens, the governing authority, the insurance carrier (if one is used), and, of course, the fire department. The plan outlines the risks in the community, what level the risks are, and whether the fire department is trained and equipped (i.e., has all the resources) to safely and effectively respond to mitigate the risks. Once fire departments determine which risks they will respond to, they then must operate in a safe and efficient manner so as not to injure or kill any fire fighters. This operational framework is commonly referred to as an "incident management system." It is incumbent upon the incident commander to determine the operational risks, and when those risks are undertaken what the benefits are—for example, persons rescued, loss stopped, property conserved. This risk-versus-benefit analysis is a constant process during incident scene operations as the incident commander makes strategic decisions based on tactical information. (For additional information see NFPA 1561, *Standard on Emergency Services Incident Management System.*)

Chapter 5, "Risk Management Planning," outlines the process of administrative risk management planning. Many fire departments have "transferred" the risk to other agencies or authorities. Although almost all communities have some level of risk associated with a potential hazardous materials incident, many fire departments have neither the training nor the resources to mitigate the incident. The U.S. government, by virtue of OSHA's CFR 1910.120, however, has determined that *all* fire departments must be trained at the operations level for hazardous materials response. Thus many departments have transferred the mitigation to either regional response teams or to private contractors. Such risk transfers are acceptable.

Chapter 6, "Role of the Company Officer and the Safety Officer," discusses the roles of these two individuals in risk management. The role of company officers or supervisors in risk management is at the tactical level. They are responsible for the supervision of personnel under their command. They must ensure that their personnel operate safely, stay together, and have the resources to do their assigned tasks. If any of these or other factors endangers the safety of the crew, they need to inform the incident commander, withdraw their personnel, and stay together. Such tactical decisions impact the overall strategic decisions, usually prompting the incident commander to recalculate the risk versus the benefit of the operation.

Concurrently, the incident safety officer functions as the eyes and ears of the incident commander in relation to the safety of operations. Safety officers also make decisions and relay information that impact operations, which again may change the complexion of the risk-versus-benefit plan.

Chapter 7, "Investigating Significant Injuries," provides some suggested guidelines on how to proceed when investigating serious injuries. The fire service focuses on the loss of fire fighters and the associated tragedy of that loss. The associated costs of injuries, including medical costs, costs resulting from time lost from work, career or volunteer shift coverage costs, and costs to those without compensation or insurance, could well total into the hundreds of millions of dollars. This area of fire fighter safety has not been looked at carefully as yet.

Fire Service Occupational Safety, Medical, and Health Issues

Stephen N. Foley

Within the fire service, the terms *safety* and *health* were not used with any degree of regularity until the mid- to late 1970s. Because of the inherent dangers of the fire-fighting profession, the assumption was that fire fighters would risk their lives doing extraordinary things to save the lives of others. Fire fighters continue to confirm the accuracy of that assumption, as evidenced by the number of fire fighter fatalities, which was significant throughout the 1970s, 1980s, and early 1990s. Since then, the number of fire fighter fatalities in the United States has hovered around 100 per year. This benchmark is still claimed by some as part of the profession. For more than a million fire fighters in the United States and the millions of others across the world, that benchmark is one that needs to be reduced.

This chapter is not intended to be a compilation of tables and figures of fire fighter fatalities and injury statistics. Instead, it is a discussion of the occupational hazards of fire fighting and what has been done and what work is currently being done to address these hazards. The primary goal of any fire service occupational safety and health program is to ensure that at the end of their shift or at the end of their career, when they leave the fire service, fire fighters can continue to enjoy a long and healthy life. Chief Alan Brunacini of Phoenix, Arizona, stated it succinctly when he said, "The fire service has suffered the most unfair occupational discrimination of any profession." It is hoped that this chapter will assist those who are either implementing or updating their department's occupational safety and health program.

Stephen N. Foley serves as the NFPA staff liaison for the Technical Committee on Fire Service Occupational Safety and Health.

Source: Section 7, Chapter 5, "Fire Service Occupational Safety, Medical, and Health Issues," *Fire Protection Handbook*, 19th Edition, Quincy, MA, 2003.

FIRE SERVICE OCCUPATIONAL SAFETY

Early Developments

Since 1987, most fire service personnel and allied professionals have referred to NFPA 1500, *Standard on Fire Department Occupational Safety and Health Program,* as either the "safety bible" or as an "umbrella document" that outlines all the components of a fire service occupational safety and health program. The development of this standard, however, was not the first venture by NFPA into fire service safety.

The initial NFPA 1501, *Standard on Fire Department Safety Officer,* was developed in the 1970s. This technical committee was dissolved by NFPA's Standards Council, and a new committee was formed in the early 1980s to begin work on a project addressing the components of an overall fire service occupational safety and health program.

The new technical committee assumed responsibility for NFPA 1501 and embarked on a course to develop NFPA 1500. The NFPA Standards Council issued the first edition of NFPA 1500 in August 1987. It was hailed by some but criticized by others as putting the fire service "out of business." Obviously this has not occurred. After a planning meeting in the summer of 1987, the committee began work on companion standards that would address those key areas in which fire fighters were found to have died or become severely injured. The key areas were incident scene management, infectious disease control, and medical requirements for fire fighters. Thus, with the expansion into specific areas within NFPA 1500, the concept of an umbrella document began to unfold.

The technical committee began revisions to NFPA 1500 in 1987, continued work on NFPA 1501 (which was later renumbered as NFPA 1521), and started documents identified in the planning session. The amount of work undertaken by this technical committee showed the resolve of those who wished to see the fire service profession as one that operated safely with a healthy workforce.

The subcommittees began work developing NFPA 1561, *Standard on Emergency Services Incident Management System;* NFPA 1581, *Standard on Fire Department Infection Control Program;* and NFPA 1582, *Standard on Medical Requirements for Fire Fighters and Information for Fire Department Physicians.* Since their initial release, these standards have taken on lives of their own and are discussed in greater detail in this chapter. Concurrently, the full technical committee was working on revisions to NFPA 1500 and 1521.

NFPA 1500, *Standard on Fire Department Occupational Safety and Health Program.*   NFPA 1500 is the umbrella standard for a fire service occupational safety and health program. The standard outlines the components of the program, including a fire department organizational statement, a fire department organizational structure, and a fire department occupational safety and health committee. Additional chapters lay out the need for and the roles and responsibilities of the incident scene safety officer and the health and safety officer, as well as the components of a fire service training and education program. The topics of still other chapters include fire apparatus, tools and equipment, protective clothing and equipment, emergency scene operations, facility safety, medical requirements, a member assistance program, and critical incident stress management (CISM).

The 1987 edition provided the initial template of a fire department occupational safety and health program; the 1992 edition was one that was fraught with controversy and dissension within the fire service. The technical committee at its ROC meeting voted not to move appendix material into the body of the standard that would address "staffing of fire apparatus." This was debated within the committee and not included as part of the revised text that was brought to the floor at the NFPA annual meeting in New Orleans, Louisiana, for adoption in 1992. The floor action was appealed to the

Standards Council and lost. This text is still in the appendix of the 1997 edition of the standard.

The next significant issue addressed by the technical committee was having sufficient personnel to conduct an "initial operation" in an atmosphere that was either potentially immediately dangerous to life and health (IDLH) or actually IDLH. OSHA defines an IDLH atmosphere as one that poses an immediate hazard to life or produces immediate irreversible debilitating effects on health. This issue, which was discussed and debated during the subsequent revision to NFPA 1500, coincided with OSHA's public review process of its revised Rules on Respiratory Protection 29 CFR 1910.134, also known as the "2 in, 2 out rule."

There are some exceptions to this, as outlined by OSHA in its regulations, but during this revision cycle of NFPA 1500, OSHA had not yet promulgated this rule (29 CFR 1910.134) and cited fire departments under its General Duty Clause 29 CFR 1910.156, while using NFPA 1500 as a reference. The citing of fire departments caused a great deal of confusion and resistance among the fire service leaders and the governing bodies they reported to. Not until the revised NFPA 1500 was passed in May 1997 and OSHA issued its rules in January 1998 was the inconsistency resolved.

The 1997 edition of NFPA 1500 included major revisions on risk management, both administratively and at emergency scenes, further discussion on the role and use of an incident management system, occupational health and wellness, and updates to references of other NFPA standards. During this 1997 revision, the technical committee revised NFPA 1521.

In 1998 the technical committee discussed the amount of work and the standards it was responsible for. The specific expertise in certain areas, especially medical and health areas, was not represented adequately within the committee structure. The technical committee petitioned the NFPA Standards Council to break into two technical committees. In January 1998, this split was approved and the Technical Committee on Fire Service Occupational Safety was assigned NFPA 1500, NFPA 1521, and NFPA 1561. The new Technical Committee on Fire Service Occupational Medical and Health was assigned NFPA 1581, NFPA 1582, and NFPA 1583, *Standard on Health-Related Fitness Programs for Fire Fighters.* Some cross-membership occurs between the two technical committees.

Currently the fire service occupational safety committee is working on revisions to NFPA 1500, 1521, and 1561, which the committee planned to bring for adoption at the 2001 NFPA fall meeting, with issuance in January 2002. The fire service occupational medical and health committee is working on revisions to NFPA 1582 and a proposed NFPA 1584 entitled *Fire Department Incident Scene Rehabilitation.* These two documents are scheduled for adoption at the 2002 NFPA annual meeting, with issuance in August 2002.

NFPA 1501/1521, *Standard for Fire Department Safety Officer.* NFPA 1501 was NFPA's first fire service safety document in the series of fire service occupational safety and health standards. Although other standards did deal with protective clothing, fire apparatus, and other issues related to the fire service, NFPA 1501 was the first to specifically address fire fighter safety and health. The first two editions of the standard addressed areas dealing with the roles and responsibilities of a fire department safety officer. These roles were many and varied and were interpreted differently from user to user. The committee struggled with how to make this document a useful tool for the fire department and how to have it play a bigger role in fire fighter safety. During the revision between the 1987 and 1992 editions, a movement began within the fire service to not only expand the roles and responsibilities of the safety officer but also delineate those of the incident safety officer (ISO) and the health safety officer (HSO). Organizations such as the Fire Department Safety Officers Association (FDSOA) advocated this process and promoted the training requirements and curriculum development in these areas. The FDSOA is accredited through the National Board of Fire Service Professional Qualifications to certify persons who meet the requirements of this standard.

While the committee was revising and expanding this standard, content experts in the field and curriculum developers at the National Fire Academy in Emmitsburg, Maryland, were developing two new courses based on the committee's work. These two 2-day courses, Fire Department Incident Safety Officer and Fire Department Health Safety Officer, outlined the basic requirements and roles and responsibilities of each position within the fire department. The course included activities that emphasized the decision-making processes each position requires and provided instruction on how to read and utilize NFPA standards, federal laws, and rules and regulations. These two courses were offered to the fire service at the time that the 1992 edition of NFPA 1521 was issued and are still being offered.

The specific roles for the incident safety officer and health safety officer within a fire department's occupational safety and health program are a critical component in making the program work. In some smaller or medium-sized departments, one person may be assigned the responsibilities of both positions, which usually leads either to that individual being overworked and frustrated or to the program not being implemented at all.

Health Safety Officer's (HSO) Roles and Responsibilities. The HSO's roles and responsibilities include the following:

- Serves as chair of the fire department occupational safety and health program
- Develops and coordinates a confidential record-keeping system
- Coordinates an accident investigation program, including apparatus crashes and fire fighter fatality and injury investigations
- Implements the department's risk management plan
- May function as the department's infection control officer or serve as liaison with the medical control personnel
- May assist in incident scene safety by either filling the role of ISO or performing other assignments as ordered by the incident commander
- Coordinates with the fire department physician
- Conducts research on apparatus, protective clothing and equipment, and other safety-related issues; may develop the specifications for these in conjunction with a representative group from the fire department
- Coordinates and/or conducts facility safety inspections

Incident Safety Officer's (ISO) Roles and Responsibilities. The ISO's roles and responsibilities include the following:

- Serves as a member of the incident commander's (IC) command staff during incident scene operations
- Develops the safety plan as part of an incident action plan
- Has the authority to terminate unsafe actions or operations at the incident scene
- Coordinates the actions of assistant safety officers as assigned at the incident scene
- Ensures that incident scene rehabilitation activities are addressed by the IC
- Assists on investigations as required by the HSO

- Ensures that a safety officer with the specificity of special operations is assigned for those specific incidents (i.e., hazardous materials or technical rescue)

The 1997 issue of NFPA 1521 has further refined the roles of each position, provided sample documents within the standard, and linked the standard to the other relevant occupational safety and health standards.

NFPA 1561, *Standard on Emergency Services Incident Management System*   The first edition of NFPA 1561 was issued in 1990. Incident scene management (or incident command, as some still call it) is a critical factor in fire fighter occupational safety and health. Since 1986, the response to and coordination of resources at hazardous materials incidents require the use of an incident command system (ICS), as dictated by OSHA 29 CFR 11910.120 and by EPA regulations. These regulations also required personnel who would be operating at a hazardous materials incident to be trained and those in supervisory positions to have additional training specific to incident command. The absence of incident command at an incident scene puts fire fighters at great risk and is one of five leading contributing factors of fire fighter fatalities, as reported by NIOSH.

This topic is one debated hotly across the fire service because the initial version of the standard went beyond the ICS and used terms that were not in that system. The discussion continued within a small group of fire service leaders on how to meld the two leading systems—ICS and fireground command—into one.

Numerous meetings, conferences, and symposiums were held across the country on how best to accomplish this task. A consortium of organizations and individuals worked to accomplish this task. The initial group included representatives from the Firescope Board of Directors, the Phoenix Fire Department, the United States Forest Service, the United States Fire Administration, NFPA, and some NFPA 1521 technical committee members (who wore numerous hats).

What emerged from the discussion is a National Fire Incident Management System Consortium that has developed six different texts that provide procedural guides of implementing incident management for different types of incident scene operations. These procedural guides compliment the NFPA standard and include references that assist both the instructor and the user.

Concurrently other national issues were developing in this area, including changes to both the National Fire Academy and the Emergency Management Institute curricula, revisions to the ICS documents as used by Firescope, and changes to fireground command. These revisions and changes include, but are not limited to, terminology, modular expansion, implementation at the incident scene, and multiagency, multijurisdictional command issues. This national movement toward compatibility is moving forward, although some would say too slowly.

The standard was expanded and refined to reflect the changes promulgated by the consortium. In addition, areas on incident scene rehabilitation, special operations, operations commanded by other agencies, additional training, and other updates to other standards were included.

In the 2000 edition, the technical committee changed the title and scope of NFPA 1561 to require all emergency services to use an incident management system (IMS). The committee made these changes because committee members realized that many agencies were present at the different types of incidents fire departments responded to. Even though fire departments were trained in and used an IMS, other agencies that lacked training and didn't use an IMS could have an impact on operational safety and the safety of fire department members. Today many of these agencies have adopted IMS and use it for training in their respective organizations.

The critical factors within an IMS are as follows:

- Establishment of command
- A strong and visible incident commander
- An incident risk management plan that includes
    - Routine evaluation of risk in all situations
    - Well-defined strategic options
    - Standard operating procedures
    - Effective training
    - Full protective clothing and equipment
    - Effective incident management and communications
    - Safety procedures and safety officers
    - Personnel accountability, by both location and function
    - Backup crews for rapid intervention
    - Adequate resources
    - Rest and rehabilitation
    - Regular reevaluation of conditions
    - Pessimistic evaluation of changing conditions
    - Experience based on previous incidents and analysis

The training in and use of incident management are paramount to the safety and survivability of fire fighters. The incident management system is a "toolbox" for the incident commander. From that toolbox, incident commanders can pick out various tools or resources to assist them in managing the incident. These resources fit into three categories: capital, personnel, and knowledge. Capital resources include local equipment and equipment identified as being available from surrounding agencies or municipalities. Personnel comprise individuals from within the initial responding agency and those from adjoining municipalities or agencies. The third category of resources—knowledge—is the most important. It consists of the knowledge base of the incident commander, as well as his or her ability to utilize the knowledge base of those involved in the incident itself. There are tools available for assistance at all types of incidents. These tools may be found in the fire department or in other agencies. The key for the incident commander is knowing how to use the tools to best manage the incident. NFPA 1561 provides the toolbox.

FIRE SERVICE OCCUPATIONAL, MEDICAL, AND HEALTH ISSUES

NFPA 1581, *Standard on Fire Department Infection Control Program.* As the fire service has expanded its role, the fastest-growing service it provides is that of emergency medical services (EMS). For fire departments that provide a level of emergency medical service, that service constitutes more than 60 to 70 percent of their responses.

The fire service provides different models and levels of this service. Included are the basic first responder level, as outlined by the medical regulatory agency; the basic emergency medical technician level (EMT-B); the intermediate level (EMT-I); and the paramedic level (EMT-P). At both the intermediate and paramedic levels, regulatory agencies may allow certain procedures based on need, training, and local medical protocol.

The models of service delivery may include treatment and transport at all levels; treatment at all levels with a combination of public, private, or third-party transport;

and treatment with only private transport. In any of these models, the fire service uses largely dual-trained cross-role fire fighters. Many departments now require members to have some level of emergency medical training and certification before they apply for a position within the department.

With these increased roles and services come increased risks. These risks within this context include communicable and infectious disease. The fire service members providing patient care place themselves at risk at an incident scene, but some of these same risks are found in the facilities in which they live and work.

NFPA 1581, 1992 edition, was part of the plan of developing standards that could have a significant and immediate impact on fire fighter health and safety. The standard outlined a program that would afford fire fighters a level of protection if they followed documented practices regarding

- Wearing protective clothing
- Being immunized and vaccinated
- Adhering to an exposure reporting system
- Cleaning and disinfecting clothing and equipment
- Acquiring needed training

Important to note is the fact that when the standard was being issued, there was an outbreak of the AIDS virus, the initial hepatitis B vaccines were being developed, and the fear of fire fighters becoming infected and possibly dying was ongoing. The federal government moved quickly, under the auspices of the Centers of Disease Control and Prevention, OSHA, and others to develop rules and regulations on protecting emergency response personnel and health care workers (Figure 3-1). The Ryan White

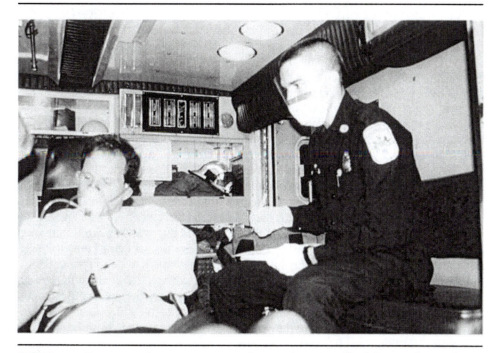

FIGURE 3-1 Emergency Response Personnel Protecting Himself While Providing Patient Care

Source: Photo courtesy of Fairfax County Fire and Rescue Department, Fairfax, VA

Act spelled out protection, notification, and confidentiality issues regarding communicable and infectious diseases.

It was incumbent upon the fire service to "clean up" its own facilities as required by the standard. This cleanup included the cleaning and disinfecting of protective clothing and equipment, a separate room and laundry facilities to accomplish that, a clean work area in the kitchen to wash and prepare meals, and clean and properly spaced living quarters.

The 1992 edition of the standard provided the initial requirements. The 1995 edition contained those additional requirements and updates on vaccinations and immunizations, facility safety, and worker protection. The 2000 edition has continued along those same avenues and expanded into other areas, including updating the CDC regulations, updating the immunization and vaccination list, and expanding the list of infectious and communicable diseases. This standard, along with NFPA 1999, *Standard on Protective Clothing for Emergency Medical Operations,* is an asset to being better educated regarding the hazards of infectious and communicable diseases.

The fire service needed an educational venue to explain to fire fighters that the protective clothing they wore at an incident scene contained contaminants that might be harmful. Many fire fighters would take home their protective clothing, let their children play in the clothing, maybe wash it with other family clothes in the home washer and dryer, and potentially have it spread the contamination to their families and friends. Or they would store this same equipment in a locker with other personal effects at the station, and then maybe wash it with the station linen, station uniforms, and the like. The possibility for such cross-contamination is real and the fire service needs to be educated about that. Even if the station does not have in-house cleaning capabilities, the fire department is required to clean the fire fighters' protective clothing every six months at the minimum. The issue is for fire fighters to protect themselves so that they can continue to provide service to others.

NFPA 1582, *Standard on Medical Requirements for Fire Fighters and Information for Fire Department Physicians.* Fire fighter medical requirements were originally included as a component of NFPA 1001, *Standard for Fire Fighter Professional Qualifications.* In 1988, the Occupational Safety and Health Technical Committee and the Professional Qualifications Technical Committee formed a working subcommittee whose mission was to write a medical requirements standard for both candidate and incumbent fire fighters. This committee, which was composed of occupational medical physicians, fire service personnel at all levels, fire service instructors, and representative fire service organizations, realized early in the process that medical evaluations for fire fighters are an important component of an overall occupational medical program.

The first edition of the standard, issued in 1992, outlined medical requirements that were categorized into A and B conditions. Category A conditions would preclude someone from either becoming a fire fighter or from continuing in a position that included primarily fire suppression. This decision caused difficulty because most fire departments had only one entry portal into the service and that was as a fire suppression fire fighter. An additional complication emerged as fire departments created "light-duty positions" or moved personnel into other positions that had no job description.

Fire department physicians or those hired to perform medical evaluations had no fire fighter job description or task analysis on which to base the medical evaluation and, thus, determine the medical condition necessary to be a fire fighter. Consequently, this raised questions on who was being "accommodated" in order to maintain his or her status as a fire fighter. Although some departments made this practice part of a labor agreement, other departments considered it part of the routine procedure. This issue led to some problems while the 1997 edition of the standard was being written.

The technical committee physicians realized that when they examined fire fighters, certain areas within the standard were critical, based on how the 1992 edition was being interpreted. They identified the following areas as ones that needed greater emphasis and clarification: cardiovascular, neurological, vision, hearing, and metabolic processes. Thus, the committee sought out specialists in those areas to assist them in rewriting the standard. Concurrently, litigation, citing NFPA 1582, was brought forward regarding the use and interpretation of the Americans with Disabilities Act (ADA). The implementation and interpretation of the ADA law created confusion on how this law impacted employment factors of hiring and promoting within the fire service. The committee worked toward revising the standard and further defining areas by developing a sample job task analysis, a sample job description, an annex that included discussion regarding ADA, and a sample physician's checklist. In the 1997 edition, all these changes were made and the medical requirements were updated.

Medical technology changes constantly; the standards-making process, however, cannot keep up with these changes. The committee members, along with outside physicians, worked within the confines of a short revision cycle to continue to refine the five identified areas in the 1997 edition. Additionally, the committee realized that the work being done by NIOSH in its fire fighter fatality investigations as well as documentation from NFPA's Fire Analysis and Research Division would assist it in the future as revisions to NFPA 1582 were being made. The 2000 edition of the standard, therefore, reflects some of those revisions, especially in the areas of diabetes, hearing, and cardiovascular disease.

Statistics tell us that almost 50 percent of fire fighter fatalities are cardiac related and almost 50 percent of those had previous cardiac-related problems. The commonsense approach would be to medically evaluate these personnel annually and try to diagnose these conditions in the early stages and treat them before they occur. Recall that the purpose of an occupational safety and health program is for the fire fighter to leave the profession in the same condition as when he or she entered it.

Fire fighting is a hazardous occupation. In addition to cardiac-related problems, other medical problems are also characteristic of the profession, including a high incidence of cancer, especially of the liver, kidney, colon, prostate, and lung. Screening for these and other occupationally related diseases is a component of an occupational medical program and is included as part of the medical evaluation program.

The fire department physician plays a significant role in managing the fire department's occupational medical program. Many fire departments that cannot afford their own physicians have grouped together to contract with either a physician or local HMO. Some smaller departments may utilize the physician as a regular member of their fire-fighting force. In any case, the physician oversees the program, reviews medical exams and evaluations, recommends follow-up or specialist exams, and works in conjunction with the health safety officer and the health fitness coordinator. In view of today's statistics regarding fire fighter fatalities and injuries, the fire department physician is an important person within an occupational safety and health program.

NFPA 1582 is moving forward with revisions that will be promulgated in May 2003. The committee is proposing developing a more stringent entry examination for candidates and a comprehensive occupational medical program for incumbents. This approach will certainly assist users and enforcers who have struggled to use the same medical requirements for both incumbent and candidate fire fighters.

NFPA 1583, *Standard on Health-Related Fitness Programs for Fire Fighters.* As part of a fire department occupational health and safety program, NFPA 1500 includes text requiring a physical performance program as well as a fitness program. The committee struggled to provide a standard on physical performance requirements that would be required for both candidates and incumbents alike. This standard was doomed from the outset. There was considerable discussion among the committee, as well as litigation within the United States, regarding the validity and content of physical performance testing. Trying to develop a set of physical test skills related by task analysis to the profession of fire fighting and then to validate that test was difficult at best.

The committee skipped a cycle and with a great deal of angst reached consensus on a document that was passed by the NFPA membership and then subsequently appealed to NFPA's Standards Council. It was returned to the technical committee by the council and withdrawn from the system after a unanimous vote from the committee members in 1997. The technical committee then returned to the drawing board in the summer of 1997.

As this proposed NFPA standard was withdrawn, a group of 10 cities within North America began work on a joint fire service wellness program. This program, supported and endorsed by the International Association of Fire Fighters (IAFF) and the International Association of Fire Chiefs (IAFC), included Phoenix, Arizona; New York, New York; Los Angeles County, California; Seattle, Washington; Calgary, Alberta, Canada; Austin, Texas; Charlotte, North Carolina; Indianapolis, Indiana; Fairfax County Fire and Rescue, Virginia; and Metro-Dade County, Florida. IAFF officials from each local, fire chiefs from each city, and fire department physicians worked on this program, which was introduced at both Fire Rescue International and the IAFF Redmond Symposium in 1999. This program has been used successfully by these 10 departments, as well as others across the world. It is a great example of a joint labor–management initiative that benefits all of the fire service.

The Fire Service Occupational Medical and Health Technical Committee began work on a health-related fitness program for the fire service, which initially did not have widespread support from the fire service. Many saw this effort as a revisit of the physical performance test versus what it really was—namely, a standard containing the components of an overall health-related fitness program for the fire service. Managed by both the fire department physician and the health fitness coordinator, this standard has nutrition, wellness, and fitness components that may be accomplished in a number of different ways. The program, as outlined in the standard, is similar to the program developed by the IAFF/IAFC Joint Labor/Management Wellness Initiative.

The 10 cities in the initiative then began work on a physical performance component that would be used to test candidates wishing to seek employment in the fire service profession. The testing process was developed with the assistance of the U.S. Department of Justice. The job tasks were validated for the 10 cities, the testing mechanism and props were developed, and then the test was run using incumbents from each department. This test, called the Candidate Performance Agility Test (CPAT), was introduced at both Fire Rescue International and the IAFF Redmond Symposium in 1999. The validity developed within the 10 cities required each to provide a job task analysis and a job description based on that analysis in order to validate the CPAT test for their municipality. The validation process is not a "one size fits all" for those who choose to utilize it. Municipalities must go through the same process as did the 10 cities and validate CPAT individually. The IAFF has developed a Peer Fitness Training Program that assists those municipalities that wish to use CPAT.

New Projects

The Fire Service Occupational Safety and Health Committee has completed a new document: NFPA 1584, *Recommended Practice on the Rehabilitation of Members Operating at Incident Scene Operations and Training Exercises*. This program, as outlined in NFPA 1500 and NFPA 1561, is part of a fire department's incident management system. The proposed document outlines the components of an incident scene rehabilitation program, including fire fighter medical evaluation, environmental protection, rehydration, and personnel accountability.

INCIDENT MANAGEMENT SYSTEM

Level of Risk

During the past 30 years or more, the U.S. fire service has evolved from a single-mission public service to one that provides a multitude of services. Inherent in the occupation of being a fire fighter is a level of risk. Each one of the services the fire service provides offers some level of risk. The level of risk has different meanings for and a different impact on individuals, departments/brigades, and the reporting authorities that have jurisdiction. This risk of injury or death is assumed by some to be an integral part of the profession and functions somewhat like a "badge of courage" within the occupation.

As the risk increases, processes have been put in place to address the risk. In incident scene operations, the risks are multifaceted, depending on the type of operation. The incident commander must assess those risks continually throughout the operation. At the incident scene, the process is one of risk versus benefit. Only recently have incident commanders been required to communicate, through an incident management system, their assessment of the risks and their strategy for managing those risks (Figure 3-2). In turn this strategy (also called goals) must have measurable tactics (also called objectives) to accomplish that strategy. Finally, the goals and objectives are put in place through an incident action plan, with supervisory officers providing input. It may sound simple, but reports show that this procedure is not followed. Laws, codes, and standards do not always provide the regulatory capabilities for enforcement; yet, to date, except in certain circumstances, the enforcement of these laws, codes, and standards carries little or no weight.

In an incident management system, members must be trained to use the system not just for large-scale incidents but for all incidents to which they respond. Under OSHA's 29 CFR 1910.120 regulations for response to hazardous material incidents, this

FIGURE 3-2   Assessing Risk as Part of the Incident Management System

Source: Photo courtesy of Fairfax County Fire and Rescue Department, Fairfax, VA

requirement has been in force since 1986. Incident management is an "all-risk" tool that has been required for hazardous materials response since 1986. If those responding agencies utilize the incident management system at those incidents, one could assume that they could make the transition of using the incident management system at "other" types of incidents.

The incident management system has a built-in risk management process based on effective supervisory levels, span-of-control procedures, standard training evolutions, standard terminology, unity of command, and sufficient resource allocation and deployment. The system, developed in the early 1970s, was used to assist with large-scale wildland fire incidents in southern California. Since then, it has developed and is used as an all-risks incident management system. If there is no system in place, the following, with sometimes tragic consequences, results: redundancy of resources, no command and control processes, no accountability, lack of a communication plan, and a high-risk environment for fire fighters. If no one has assumed command of the incident, then usually many think they are in command. Multiple commands mean multiple plans, with no one in charge—a truly risky environment from which to operate.

The incident management system is designed to expand modularly to effectively create supervisory positions so a span of control of three to seven positions per supervisory level is maintained. This span of control gives the system the capabilities of effectively managing resources—that is, personnel.

Accountability

Crew/company supervisors must keep track of the personnel assigned to them. In conjunction with that requirement is a personnel tracking system, assigned by the incident commander, to track personnel both by location and function. Therefore, if there is an incident management system in place, accountability is effectively accomplished. This process keeps the crew/company intact, or, as it is termed, "crew integrity" is maintained. So as individuals are assigned as a crew, they leave and enter as a crew. Some may think of accountability as just a process to count heads if an area or building is being evacuated. Although that is a component of the accountability process, it is much more than that. The accountability process also includes rehabilitation, air supply replenishment, and relief assignments.

Communication

Lack of or ineffective communication usually leads to a multitude of problems. If incident commanders cannot communicate with their personnel, they are out of business! Communication is integral to incident scene operations; it is the key to fire fighter safety and survival (Figure 3-3). If the premise of incident command is a defined strategy with tactical objectives, then the person in command must be able to communicate with supervisory personnel. If supervisors need additional resources—that is equipment or personnel to complete an objective—then they need to communicate with the incident commander. If the resources cannot be provided and the tactical objective cannot be accomplished, strategy is impacted. If the strategy and the type of operation are changed, based on new information and a revised risk analysis, everyone needs to know what those changes are. Communication is also a component of the accountability process. The ability to communicate with supervisors overseeing a specific function or in a geographic location is part of the accountability process.

FIGURE 3-3   Communication as an Integral Part of Accountability
Source: Photo courtesy of Phoenix Fire Department, Phoenix, AZ

In addition, part of the communication process includes the use of standard terminology. If a crew is assigned to a specific location, for example, Division 4, and the crew requests additional personnel, do the additional personnel know what and where Division 4 is? National standards require the use of clear text in communication transmissions. Slang, local jargon, amateur radio text, ten codes, and other colloquialisms only hamper the communication process. Again, standard terminology is a component of the incident management system. Agencies that operate within an agency management system should be trained in the proper communication protocols.

Standard Operating Procedures

The incident management system and all of the components within it require a fire department to have standard operating procedures (SOP) in the training and use of the system. Some fire departments are under the illusion that a standard operating guideline is the same as a SOP. SOPs are requirements, not guidelines. If a department uses and enforces its SOPs, then the incident commander has a level of confidence in the use of the system and assignments at the scene. Many departments use their SOPs as part of training, part of promotional procedures, and as part of a post-incident analysis tool.

ON-DUTY FIRE FIGHTER DEATHS AND INJURIES

Since 1977, NFPA has documented more than 3200 fire fighter fatalities in the United States that resulted from injuries or illnesses while the victims were on duty. The NFPA also conducts an annual survey of fire departments that results in estimates of the number of nonfatal on-duty injuries that occur each year. The following tables summarize the data: Table 3-1 shows the distribution of on-duty fire fighter deaths from 1977 through 2001; Table 3-2 shows the distribution of deaths of local or municipal fire fighters over the same period; Table 3-3 provides a breakdown of on-duty deaths by age and cause of death in 2001; and Table 3-4 shows the distribution of on-duty fire fighter injuries from 1981 through 2001. The 2001 figures do not include the 340 deaths at the World Trade Center.

TABLE 3-1　On-Duty Fire Fighter Deaths, 1977–2001

Year	Deaths	Year	Deaths
1977	157	1990	107
1978	172	1991	108
1979	125	1992	75
1980	138	1993	79
1981	136	1994	104
1982	127	1995	97
1983	113	1996	96
1984	119	1997	98
1985	128	1998	91
1986	120	1999	112
1987	131	2000	103
1988	136	2001	99
1989	118		

TABLE 3-2　Career versus Volunteer Fire Fighter Deaths—Local or Municipal Only, 1977–2001

Year	Career	Volunteer	Year	Career	Volunteer
1977	70	82	1990	25	62
1978	64	100	1991	36	66
1979	58	57	1992	24	44
1980	61	69	1993	21	55
1981	58	65	1994	33	38
1982	50	67	1995	29	59
1983	54	51	1996	27	65
1984	43	59	1997	31	58
1985	55	66	1998	33	49
1986	51	55	1999	38	70
1987	48	68	2000	28	58
1988	43	81	2001	24	63
1989	43	65			

TABLE 3-3　On-Duty Fire Fighter Deaths by Age and Cause of Death, 2001

Age Group	Heart Attack	Non-Heart Attack
20 and under	0	3
21–25	0	7
26–30	2	8
31–35	3	2
36–40	4	12
41–45	5	6
46–50	3	10
51–55	9	4
56–60	2	3
over 60	12	4

TABLE 3-4　On-Duty Fire Fighter Injuries, 1981–2000

Year	Injuries	Year	Injuries
1981	103,340	1991	103,300
1982	98,150	1992	97,700
1983	103,150	1993	101,500
1984	102,300	1994	95,400
1985	100,900	1995	94,500
1986	96,450	1996	87,150
1987	102,600	1997	85,400
1988	102,900	1998	87,500
1989	100,700	1999	88,500
1990	100,300	2000	84,550

NATIONAL INSTITUTE FOR OCCUPATIONAL SAFETY AND HEALTH (NIOSH)

NIOSH's Role in Fire Service Occupational Safety

The National Institute for Occupational Safety and Health (NIOSH), an agency within the Centers for Disease Control and Prevention, received authorization and appropriations from Congress in 1998 to investigate all fire fighter fatalities within the United States. This investigation process includes career, volunteer, military, and federal fire fighters. As part of the investigation and reporting process, NIOSH posts its investigation reports on its website (www.cdc.gov/niosh/firehome.html).

As reported by NIOSH, the contributing factors to fire fighter fatalities on the fireground in the United States are the following:

1. Lack of incident command/management system
2. Inadequate risk assessment
3. Lack of fire fighter accountability
4. Inadequate communications
5. Lack of standard operating procedures

In fact, items 2 through 5 are integral to having and using an incident command system.

NIOSH's Cardiovascular Disease Investigation

Part of the NIOSH investigation is the study of fire fighters who die as the result of cardiovascular disease. Each year approximately 100 fire fighters are killed in the line of duty. Cardiovascular disease (CVD) is the number-one cause of these on-duty fatalities, typically taking 45 fire fighter lives per year. CVD is not only an occupational health problem afflicting the fire service but is also a public health problem. To address this problem, NIOSH conducts fatality investigations of on-duty fire fighters killed as the result of CVD.

The investigation includes an assessment of the physiological and psychological demands of the job, workplace organizational factors (screening tests), and an assessment of individual risk factors for coronary artery disease. Each investigation culminates in a succinct report distributed to the affected fire department as well as the country's fire service. Circumstances of each fatality are entered into a database for analysis. These investigations and subsequent analysis of the database provide insights for prevention and intervention activities.

Objectives and Evaluation of NIOSH's Program

NIOSH has formed two-member teams utilizing subject matter experts from NIOSH as well as physicians to conduct the CVD investigations. The program goal is for each team to conduct 12 investigations of CVD and other causes per year. In addition, members of these teams are responsible for publishing peer-reviewed journal articles and giving presentations at national meetings. The following investigations have been completed.

- 1998: 10 investigations
- 1999: 21 investigations
- 2000: 23 investigations

SUMMARY

NIOSH is providing a valuable service in assisting the fire service in determining causes of fire fighter fatalities. The information gathered is being used to create and revise NFPA standards, educate allied professionals on the hazards of the fire-fighting profession, and, most important, reduce the number of fire fighter fatalities.

Evaluation and Planning of Public Fire Protection

Revised by John Granito

The purposes of this chapter are to demonstrate that adequate fire protection and related public safety services are essential components of any community and that both the level and types of services provided should be consciously and carefully considered choices of the community that are made with, and based on, adequate information. Further, objective evaluation and planning can ensure that a community's desire for a stipulated level of protection is being met in as cost-effective a manner as possible. Several other chapters provide information useful in the self-assessment of a public fire protection system. In addition, certain NFPA publications, including a number of codes, standards, and recommended practices, are applicable.

BACKGROUND ISSUES CONCERNING FIRE PROTECTION

Because the safety of people, property, and the environment is so vital, questions that focus on the organization and deployment of fire-fighting resources are especially important to communities. Prompted by escalating costs, issues relating to fire suppression have attracted a great deal of public attention, even though fire fighting is but one part of what constitutes community fire protection. Although fire prevention and fire protection engineering efforts are typically less costly in the long run than fire fighting, questions relating to the appropriate size, deployment, and response protocols of fire-fighting forces remain paramount.

The first issue is created by the relatively high cost of fire stations, fire apparatus and equipment, and on-duty personnel. A second set of issues is technical in nature, dealing with the number of responders necessary to be on duty and to be dispatched, with what should be the maximum allowable time for response, and with other related technical and operational questions.

John Granito, Ed.D., is a consultant in fire-rescue services. He is professor emeritus and former vice president for Public Service and External Affairs, State University of New York, Binghamton, New York.

Source: Section 7, Chapter 2, "Evaluation and Planning," *Fire Protection Handbook*, 19th edition, NFPA, Quincy, MA, 2003.

The third issue considers what the process should be for determining the answers to the technical and operational questions.

- Should communities assess their own needs and resources and provide individualized answers, or should national standards be applied?
- How should national standards and recommended practices be formulated?
- Should noncompliance with a national or industry standard expose officials to penalties?
- Should fire chiefs be obligated to inform fully their local officials concerning local hazards, risk levels, and levels of service actually available?
- Should local officials keep citizens fully informed concerning available emergency service capabilities?

These and similar questions have generated much debate about the process of evaluating community fire and rescue defenses.

Unfortunately, it appears possible that even those who are genuinely concerned with evaluating and planning for local fire, rescue, and emergency medical services may blend together the three issues of cost, technical correctness, and responsibility for resource decisions. When this happens, communities cannot be sure that the resulting delivery system is the most appropriate obtainable for the available financial resources.

Of course, there are multiple demands for local municipal funds. Thus, emergency protection must be fitted into a larger community priority listing of needs and desires. However, proper placement in the listing requires that a true and accurate description be publicized of existing capability compared with accurately described community hazards, response history, and call demand predictions. When the three issues are erroneously blended together, citizens may believe that adequate, timely, and safe response is available when indeed it is not, and may give higher budget priority to another community interest. Or citizens may be unduly anxious over erroneous observations that fire defenses are much too meager and that even more resources must be allocated.

Citizens are often subjected to opposing budget arguments concerning protection levels and frequently lack the knowledge necessary to arrive at sound conclusions. This understandable lack of technical expertise underscores the necessity for accurate, well-balanced, and forthright information to be widely disseminated among those who will both receive and pay for service delivery.

Partly because the benefits of active comprehensive prevention programs, public safety education, and fire protection engineering efforts are not well advertised, there often is little pressure exerted on their behalf. Thus, relatively few resources are allocated to these efforts in many communities when compared with allocations for suppression work. Even though it certainly is true that the least harmful fires are those that never occur, most communities fail to consider a total fire protection plan as a wise and cost-saving instrument. The benefits of residential and high-rise fire sprinklers are well documented, for example, yet conflicting pressures have often curtailed their use. The funds allocated to fight the fires that do start, however, can be quite high. Much more emphasis is needed on the total protection program if communities are to be made reasonably safe at reasonable costs. The evaluation and planning of public fire protection, then, must consider total protection efforts. These certainly include provisions for emergency response, but also such efforts as locally adopted codes and ordinances, built-in structural protection, inspection programs, public education for all groups, building plan reviews, and so on must be included as well.

The same concept, of course, applies both to the preventive and emergency-type pre-hospital medical services fire departments often provide.

A practical approach to the development of a community's mid- to long-range plan is to first generate a view of what the community will be by the time of the planning horizon. That information comes from an analysis of past development combined with projections of demographic changes, economic development or decline, major roadway programs, annexations, significant building projects, and so on.

The second step is to assess realistically the strengths and weaknesses of the existing fire protection system, including codes, standards, and ordinances relating to safety, fire prevention efforts, public safety education programs, and emergency response capability.

The latter subsystem, that of response strength, involves an honest assessment of stations and their locations, vehicles and equipment, staffing numbers and provisions, breadth and depth of the various service delivery areas, training and other support activities, success rate in handling emergencies, mutual-aid arrangements, and dispatch and communication provisions.

In 2001 the NFPA consensus standard building process produced two new standards, one for substantially career fire departments (NFPA 1710) and the other for substantially volunteer departments (NFPA 1720).

NFPA 1710, *Standard for the Organization and Deployment of Fire Suppression Operations, Emergency Medical Operations, and Special Operations to the Public by Career Fire Departments*, deals with the time frame for response to fire and medical incidents for the first-due engine company; for the entire first alarm assignment; and for the first level, basic, and advanced emergency medical responders. The number of fire fighters necessary for initial suppression attack and their roles are specified in the standard, as are other aspects of organization, suppression, operations, emergency medical and other responses, special services, staffing, and so on.

NFPA 1720, *Standard for the Organization and Deployment of Fire Suppression Operations, Emergency Medical Operations, and Special Operations to the Public by Volunteer Fire Departments*, addresses the organization, operation, and deployment of those departments that are substantially volunteer in staffing. Although not as specific as is NFPA 1710 in the number of responders required or in the maximum response time frame, the standard does call for prompt initiation of suppression attack and the presence of a sufficient number of personnel for sustained attack.

The third step is to project the needed capabilities and capacity of the fire protection system and its vitally important fire department component as the community changes. Any additional needs of the system and the fire department, so that they can meet the projected demands, represent the content components of the plan. How that plan may be implemented and the time and timing required constitute the strategic and tactical planning necessary for success.

A major challenge in this "discovery" process often lies in assessing the existing protection system and its fire department and in determining the steps needed to bring both to the strength required by projected demands. Additionally, tests of efficiency, cost-effectiveness, economies of scale, and other sound business practices must be applied.

The basic questions to be raised concerning this entire process are:

- What types and levels of protection does this community need both now and to meet the upcoming community profile?
- How are the details of these needs best determined?
- What process and whose expertise should be used to arrive at the answers?

- How may the correctness of the answers be tested?
- Are the answers as cost-effective as possible?
- How can any needed improvements best be funded?
- What should be done during and after the planning process to win the support of those directly affected by changes and those who will have to pay for any improvement?

STEPS IN PLANNING

Public fire protection, emergency medical services, and technical rescue provisions need to be carefully planned and require that certain logical steps be taken to achieve a comprehensive, acceptable, and workable plan. There are two important aspects to any good plan: (1) the plan itself, which must be feasible and directed toward clear goals; and (2) the process by which the plan is developed, which must ensure that all major goals are considered and every constituency to be affected by the plan is reasonably involved in the planning process or made well aware of its consequences prior to plan approval.

Without adequate involvement of the necessary constituencies, implementation of the plan may likely fail because of a lack of cooperation and commitment. For a satisfactory plan to evolve, the planners must decide the end results they wish to achieve (goals), determine the status of the fire protection system in relation to those goals (evaluation), and calculate how much and what kinds of progress actually can take place over a certain period of time (objectives, tactics, time frame). Each of these three steps—setting goals, evaluating, working out the details—requires the collection and analysis of relevant information. Broad goals are achieved through planned strategies and precise objectives are achieved through implemented tactics. Each must be relevant to the other.

Careful consideration of these important factors, which will vary considerably from one community to another and which must take into account assets, liabilities, challenges, and available resources leads to what is often referred to as a *strategic plan.*

Because some degree of public fire protection is almost always in place, it is common for the entire process to begin with an evaluation of the fire protection that is already available. The information obtained from the evaluation, when analyzed in terms of broad, generally recognized public fire protection goals, identifies needs and provides responsible community officials with the approximate parameters of the plan to be developed.

It is important to take into consideration those fire protection and other public safety services that may be provided by a fire/rescue department and that will meet specific local needs. Examples of this are hospital transportation by a fire department ambulance and responsibility for emergency management.

Typical fire department services include fire prevention programs—such as code administration and enforcement through plans review, inspection services, and public education—as well as fire suppression, technical rescue, emergency medical, hazardous materials response, and disaster response. Various other public services, such as health screening programs, are conducted more and more often by progressive departments. This approach to broad-based community safety service takes advantage of departmental resources, such as neighborhood fire stations and on-duty personnel, experience in public service and education, and the typically high regard in which fire departments are held by citizens.

As already noted, the plan cannot be developed without the involvement of a wide variety of community groups. Even if the various constituencies seem willing to allow fire protection officials to develop and write the comprehensive plan without their consultation, the fire officials should be cautious; the citizens eventually must be willing to accommodate and pay for the implementation. Since a comprehensive plan envisions a larger group or system of integrated parts, a number of organizations and agencies outside the fire department will need to play important roles in implementation of the plan. Local elected officials typically play key roles in plan approval, implementation, and funding.

Modern practices make it necessary for fire departments to work cooperatively with a variety of agencies and organizations. These range, for example, from the local building code enforcement staff to the U.S. Coast Guard for hazardous materials spills near navigable waterways. Good planning requires consultation with all operational components on a continuing basis. A good plan today is better than a perfect plan that might be developed tomorrow.

As noted, one basic aspect of a comprehensive public fire protection plan is the concept that it is infinitely better for a community to prevent fires altogether, or to mitigate them automatically through fire safety education and built-in fire protection features, than to depend solely on the fire suppression capabilities of the community's fire department. The goal of reducing the incidence and effects of fire involves all aspects of fire prevention. Historically, much more energy and many more resources have been devoted to evaluating, planning, and implementing fire suppression/firefighting capabilities than fire prevention capabilities. Simply stated, the United States as a whole has focused more on "fire engines and fire fighters" than on public awareness and built-in mitigation features. That focus is still necessary and exceedingly important, but effort must also be placed on measures that do not depend solely on fire suppression personnel to reduce the toll of fire. Planning groups have difficulty, however, in evaluating the degree of effectiveness of such multifaceted fire protection programs.

A reasonably effective method exists for conducting the evaluation, involving three kinds of analysis. The first kind of analysis requires the community, typically through the fire department, to maintain consistent and carefully recorded information each year concerning the number of fires that occur and the cost of those fires in lives and dollars, including wages and tax revenues lost. When those human and physical costs are added to the expense of maintaining fire prevention and suppression systems, plus the cost of fire insurance, a standardized total cost of fire to the community can be compared with that same cost in earlier years, or over an average of three or more years. Necessary adjustments for inflation, community growth or decline, and other important variables can be made by local officials, thus permitting a community to compare its present to its prior fire performance.

The second kind of analysis most useful in determining cost-effectiveness involves communities identifying other communities that are similar in ways important to fire protection (such as services desired, size, construction, hazards, and geography) and comparing their own total performance to the total performance of the similar communities. This process is termed *benchmarking* and has become a popular methodology. However, the number of site-specific variables may be so high and the span of differences so great that comparisons prove confusing. Benchmarking requires cautious application.

The third kind of analysis, although self-conducted, is formatted and guided by the fire department accreditation process now available in the United States.

In the first analysis, the community uses an internal data source (itself) as a yardstick, and in the second analysis it uses an external yardstick (other communities). As more data are collected concerning the total cost of fires in various communities operating with various public fire protection plans, any given community will be able to benefit relatively quickly not only from its own experience, but also from the experiences of others.

The Commission on Fire Accreditation, International, which was developed by the International Association of Fire Chiefs in conjunction with other national organizations, provides a standardized methodology for fire departments to assess and rate themselves—with the assistance of a visiting team—along a series of categories important to community protection. Performance indications serve as points of measurement and are important components of this accreditation process. More and more departments are using this process to evaluate their condition and to benefit from both the process and the accreditation status.

Important to this planning process is the ability and willingness of the community to finance and otherwise support the total level of fire protection required by the plan's goals. Examples include legislation dealing with the retrofitting of sprinkler systems, or innovative thinking that produces a core group of senior citizen volunteers who conduct fire safety programs for their peers. In some states built-in fire protection requirements are set at the state level, and local government may not increase the level of fire safety provided by building features (automatic fire sprinklers, noncombustible roofing, etc.).

Although fire protection officials must always be concerned with reducing the total cost of fire (fire loss, plus costs of insurance, prevention, and suppression), citizens living in tight economic times will ultimately reserve the right to make decisions, or trade-offs, concerning the level of protection they wish their tax dollars to purchase. However, the safety of emergency responders cannot be reduced as part of a cost-reduction program.

To assist citizens in making decisions concerning budgets, fire protection officials must accurately describe the effect on total cost if additional or fewer resources are applied to particular prevention or suppression efforts. This kind of technical knowledge and analytical ability of officials provides a crucial element to comprehensive planning and evaluation. It is one of the most important responsibilities of the public fire protection officer.

One of the most serious issues faced by public officials and residents is that of adequate fire department staffing. In communities served by volunteer fire fighters, there may be serious volunteer recruitment and retention problems or problems in providing timely response during working hours when many of the volunteers may not be in the community they protect. In communities served by career, full-time fire fighters, restricted budgets often limit the number of fire-fighting vehicles ready to respond, or the crew size for each vehicle and, thus, the total number of trained personnel responding to the incident in time to mount an effective offense against the fire, often termed "initial attack." Fire protection officers and municipal officials need to communicate objectively the level of service being provided and how changes in fire department resources will affect that level of service. That is, the type and level of risk should be expressed in terms understandable to the fire protection nonexpert and neither over- nor under-stated. Of course, the fire department must bear the responsibility of formulating and operating a cost-effective organization, no matter what level of service is desired.

EVALUATION

In addition to assessing the capabilities of the fire department and related aspects of fire protection, fire officials evaluating the capability of the existing fire protection system must take a number of factors into account. Examples include known combustibles, the life hazard that exists, fire frequency, climatic conditions, demographic and geographic factors, and a basic consideration of the specific role of a public fire department in providing fire protection to the community. Failure to consider each of these factors adequately can lead to a large-loss fire. A fire department's suppression capabilities can never be expected to compensate totally for the deficiency or lack of built-in fire protection systems.

The evaluation of local fire suppression capability, although no more important than the evaluation of local prevention and public safety education efforts, is a necessary effort for two primary reasons. First, most communities have little understanding of either the level of local protection available, or the level necessary for reasonable community well-being. Second, the time availability of volunteers is quite limited, and the cost of full-time career personnel typically is significant. The actual number of available crew members—while exceedingly important—may not be realized by non-fire officials and citizens. This lack of information concerning actual resources can lead to a false sense of security.

Two concepts are useful in local suppression considerations. First is the "capability" of the fire department to respond within a short time with sufficient trained personnel and equipment to rescue any trapped occupants and confine the fire to the room or building of origin on initial attack. How many crew members and vehicles of various types are necessary depends on the type of call and the conditions of the local community, which either aid or hinder fire fighting. These variables, which certainly number more than 20, range from water supply and sprinkler ordinances to weather conditions and the age of buildings.

The second concept is that of "capacity," which is the ability of the fire department to respond adequately to multiple-alarm incidents ("sustained attacks") and/or simultaneous calls of any type, including emergency medical responses. If alarm patterns are examined, the volume of multiple alarms and simultaneous response demands over a period of time can be approximated. Larger municipalities typically average more demand for capacity and, thus, typically have larger departments; but, obviously, remaining capacity is diminished as suppression units are deployed, even in the largest departments. It must be recognized that although a few fireground tasks may be performed sequentially, most need to be performed consecutively, and these latter tasks require more personnel.

In evaluating both local response capability and capacity, officials need to consider the following: In most areas it is relatively easy to increase capacity (through the use of mutual aid and group or entire shift callbacks), but it is much more difficult to improve capability, which requires *immediate* response of nearby forces. A common rule of thumb is that a community using on-duty crews at fire stations should be able to have an initial attack team composed of an entire first-alarm response on the scene within approximately 10 min of receipt of the alarm. This equates to about 8 min of running time.

Another rule of thumb is that, for a totally volunteer, nonstaffed station, the average time from when a call is received to the moment a crewed vehicle leaves the station can be about 6 min. In those situations, then, the "call to on-scene time" for a 6-min run is about 12 min for one or more vehicles.

Several techniques are used by progressive fire officials to help mitigate these handicaps to achieving adequate response capability and capacity. These include using *automatic*, immediate mutual-aid response from nearby departments, rapid callback systems, volunteer personnel who "bunk in" at stations, part-time on-duty crew members, special shift arrangements, elaborate regional mutual-aid agreements, and other methods. "Doing more with less" is a popular concept, but there are limits that must be realized.

As part of a local evaluation, fire officers and other community officials can determine the status of their response forces by establishing what capabilities and capacities are necessary for adequate protection of their area and then determining what is available.

Following a local hazard analysis, for example, the following types of "what should be" questions concerning *what is needed for adequate response capability* may be posed:

- Maximum time for call processing at dispatch center?
- Minimum number of persons on first-out pumper?
- Minimum number of persons on first-out aerial?
- Minimum number of persons on first-out heavy rescue?
- Minimum number of persons on first-out emergency medical service (EMS) vehicle?
- Weekday get-out time for first-due unit?
- Weekday get-out time for second-due unit?
- Nighttime get-out time for first-due unit?
- Nighttime get-out time for second-due unit?
- Maximum clear weather running time (pumper) in district?
- Maximum clear weather running time (aerial) in district?
- Maximum clear weather running time (EMS) in district?
- Number of qualified fire fighters plus incident commander to be at scene within 8 to 10 min of dispatch?
- Number of qualified fire fighters plus incident commander to be at scene within 15 min?
- Number of minutes for mutual aid to arrive at scene?

Figure 4-1 illustrates one method for listing the actual response capability of a department for a simple, single-family-dwelling fire. Figure 4-1 can be expanded to include first-alarm responses to other types of occupancies. Response times can be verified and listed for the components of the first-alarm assignment, and other questions can be posed to identify local response capacity to handle multiple and/or simultaneous alarms.

An important stipulation put into effect in 1995, and continuing beyond that year, is that a minimum of four trained personnel be at the scene of a structure fire before an interior attack can be launched, except under the most extenuating circumstances. Both the U.S. Occupational Safety and Health Administration (OSHA) and NFPA have issued documents dealing with this. Also, in the United States and in certain other countries, various governmental agencies have issued regulations and advisories concerning the necessity for certain emergency operations to have trained incident commanders and standby safety/rescue crews present, as well as to have a defined incident command system operating. Local officials are well advised to have current information concerning all applicable federal (national), state, and provincial requirements, as well as applicable NFPA codes, standards, and recommended practices, as issued from time to time.

Rural Fire Protection

One principal difference between operation of rural and urban fire departments is that rural departments must deal with water supply issues with a broader variety of solutions than most urban departments. Rural fire department operations and apparatus emphasize not only fire-fighting requirements but also the provision of water for fire fighting. Rural fire apparatus must have large water tanks to permit effective initial at-

Number of Resources Dispatched	Confirmed Working Single-Family-Dwelling Fire
Chief(s)	
Aides or incident command technicians	
Company officers	
Fire fighters, including "flying squads"	
Single-purpose emergency medical technicians (EMTs)	
Standard pumpers	
Quints	
Aerials	
Ladder tenders	
Ambulances	
Heavy (technical/urban) rescues	
Light/medium rescues	
Mini/midi attack pumpers	
Special operations (air, lighting, etc.) units	
"Flying squad" vehicles	
Mobile command vehicles	
Safety officer	
Rapid intervention team	
Other (types)	

FIGURE 4-1   Typical, Actual, First-Alarm Fire Attack Assignment

tack on fires while supplementary water supplies are being brought into action. Supplementary water supplies include drafting sources on or adjacent to rural properties, and mobile water tanker vehicles for transporting water from more distant sources. Also in use are portable folding canvas tanks into which tank trucks quickly discharge their water supply through special dump valves. Rural fire departments often use apparatus and hose to relay water from sources several thousand feet (1000 ft equals

304.8 m) from the emergency. Initial response of pumpers, tankers, and auxiliary apparatus should be adequate for a quick attack on the burning property. With adequate highways and well-designed apparatus, it is often possible to bring substantial firefighting forces to an emergency in rural areas in sufficient time for a properly planned and executed initial fire attack operation to be effective, even though many of the personnel may arrive in private automobiles. Many rural properties are now located in areas that enjoy some level of fire protection. Some properties, of course, may have to depend entirely on their own private fire protection and whatever help they may obtain from forestry agencies or distant fire departments.

Newer approaches to fire insurance rating give "protected" status to property without a municipal water supply system if a fire station is within five travel miles and a stipulated gallons per minute flow can be maintained by the local department.

Minimum protection for a rural area would include a pumper with a large water tank and a water tank vehicle responding on an initial alarm. Properly designed tanks should be able to transport water from a source 1 mi (1.6 km) from the scene so a minimum of 250 gpm (946.25 L/min) can be pumped at the fire scene by the pumper. Since a larger flow is often required to provide adequate fire protection services, additional tankers must be used, or drafting sources within reasonable distance of the fire scene must be identified. Programs that encourage the construction of year-round rural drafting sites are to be encouraged. Rural apparatus should carry 3½-in. (89-mm) or larger supply hose to provide adequate water supply at the fire scene. It is always advisable to lay large-diameter fire hose from the water supply source to as near the fire scene as possible in order to avoid extensive friction loss. At the emergency, large-diameter hose is sometimes connected into smaller handlines, or used more often to supply another pumper from which handlines are extended. Other pieces of equipment, such as rescue and aerial ladder vehicles, should be provided as needed to carry out the mission. Elevated master streams are not needed extensively in rural operations, and sometimes ladder truck equipment for rescue, forcible entry, ventilation, and salvage operations is carried on pumpers and equipment vehicles.

To be even minimally effective in controlling a fire, the initial responding apparatus should reach the emergency scene before very rapid fire spread. As is the case in urban fire fighting, this is termed "initial attack" and is aimed at stopping the fire as close to the point of origin as possible. So-called "sustained attack" attempts to reduce the loss to the exposed adjoining or nearby property. Because of longer response times, rural departments may find themselves in the sustained attack, or "defensive," mode upon arrival. Unless sufficient water can be made available within a short time frame, the British thermal units (Btu) generated by the burning material cannot be absorbed so the temperature is not reduced sufficiently to extinguish the fire. The keys to successful rural protection planning usually reduce to response times and water availability.

Of special concern are the sometimes extensive supplies of fertilizer and pesticides located on rural properties. Response crews must be alert also to the possibility of above- and below-ground storage tanks for fuel and other products. The safe operations guidelines for hazardous materials response are applicable in rural as well as urban areas.

Another concern for rural fire protection exists because so many large and diverse types of structures are located in rural areas. These include warehouses, truck centers, product distribution facilities, processing plants, storage buildings, centralized schools, churches, trailer parks, and others.

In addition, the extension of existing housing developments and the construction of new residential and commercial structures immediately adjacent to wildland areas, or in the midst of such areas, have brought about a significant increase in wildland–urban interface fires, which present new challenges to both urban and rural fire departments.

However, it is possible to give such counsel if one can make certain assumptions that may be controlled by the community. These assumptions relate primarily to the presence of appropriately placed smoke detectors and other types of fire detection devices and a means to relay that alarm to the fire suppression agency, but they also include the existence of enforced building codes that provide some degree of resistance to fire spread from the room of origin, rural sprinkler systems, home escape plans, and so on.

Urban Fire Protection

In urban areas, inadequate fire department response to initial alarms can be a major factor in fire losses due to high population and structural densities. The number of simultaneous fire-fighting operations that may need to be conducted at the incident also dictates the total amount of personnel and equipment needed to provide effective fire-fighting operations. In any "working" structural fire, several operations must be carried on simultaneously and the fire attack must be made from several points. This cannot be accomplished by the crew of a single fire apparatus. Multiple apparatus must be positioned properly, and adequate waterflow made available to cope with the amount of fuel (fire load) involved or exposed.

In simplest terms, structural fire suppression in an urban setting involves the accomplishment of at least the following tasks, many of which must occur almost simultaneously to ensure effective and safe operations (the proper sequence will vary, depending on circumstances, as will additional tasks):

- Command of the incident (to ensure both effectiveness and safety of the fire fighters)
- Application of water in appropriate quantities (dependent on the fire environment and other factors)
- Provision of appropriate source of water supply for above
- Ventilation of smoke and other hazardous products of combustion from the fire area to the outside
- Search for and rescue of fire victims
- Forcible entry
- Control of utilities
- Salvage and other property conservation operations
- Standby rapid intervention/fire fighter rescue

At large-structure fires, additional fire-fighting personnel are needed to cover the various points of fire attack. In some cases, various functions can be handled more efficiently by specially trained crews such as rescue companies and hazardous material teams operating from specially equipped apparatus.

The number of personnel and equipment necessary to accomplish the above will vary with a number of factors [i.e., the expectations placed on the mobile suppression group, the material burning, the construction of the building, the type of built-in protection provided, separation between buildings, availability of water supply, the number and physical and emotional condition of persons in the fire building, the type of equipment available to fire fighters, the level of proficiency of the fire suppression crew (including the commanders), etc.]. Hence, it is difficult to determine a minimum number of fire fighters or equipment required without careful, objective planning, and without considering the important variables. Obviously, personnel needs will differ significantly between a small detached structure fire and a high-rise fire. Pre-incident planning is essential.

Some general considerations may, however, be listed:

- The more arduous the expectations placed on the mobile fire suppression crew, the greater the required resources (e.g., the community that expects its fire department to contain fires to the room of origin should expect to provide more fire suppression resources than the community that expects its fire department only to prevent the spread of fire from one building to another. Given the same level of protection demands, the community that leaves all fire protection to the mobile fire suppression force will require a more extensive suppression force than the community that requires a high level of built-in protection).

- The more extensive the concentrated fire potential, the greater the required fire suppression resources (e.g., given the same expectations of its mobile fire suppression force, a community having high-rise buildings, a high population density, and extensive industrial risks will normally require greater fire suppression resources than a largely residential community).

- The broader the services provided by a fire protection agency, the greater the need for resources (e.g., a fire agency providing emergency medical services will, given the same level of expectations for its mobile suppression forces, require more resources than an agency providing only fire protection services, assuming a significantly increased *total* workload demand, including a significant increase in simultaneous calls).

- The greater the geographic area protected, the greater the resource requirement of the mobile fire suppression forces (i.e., given the same service level expectations, a community providing service to 20 sq mi (52 km^2) will require more resources than a community providing protection to only 10 sq mi (26 km^2); this is caused by the need for timely arrival at the scene of fire, medical, or environmentally threatening incidents). Computerized geographic information systems and accompanying computerized response maps have powerful ability to demonstrate visually the response effectiveness of various station locations.

In most smaller and medium-size communities, all initial-response (first-alarm) apparatus will not arrive at the fire scene simultaneously. In many departments with on-duty personnel, apparatus has to respond from more than one station, and some apparatus have longer travel times to the fire scene. In volunteer departments, personnel must travel varying distances to get to the fire station or the fire scene, and, thus, all apparatus cannot go into operation at the same time. Those fire fighters and vehicles that cannot arrive at the fire scene within the first critical time period have limited impact on the initial attack, regardless of the department's response assignment ("running card"). Communities may have a false sense of security in this regard, until actual response times are tested and working initial-attack personnel counted.

The critical numbers for policy makers to use in planning related to staffing of shifts and apparatus crews—sometimes termed "minimum manning"—are those that describe how many trained personnel can arrive within a stipulated initial attack time frame. Ideally, all or most will arrive as composite crews, and not as individuals.

The total minimum fire force recommended for any community is necessarily dependent on community hazards and the expectations of the community and the members of the mobile fire force. Objective planning and evaluation are necessary to determine the resources required for effective and safe fire suppression operations that will meet community requirements.

Some fire agencies are addressing the suppression challenge with different arrangements of fire crews within the communities than were heretofore generally accepted (e.g., use of multipurpose apparatus task force assignments, and differential staffing). It is important to note that, although these different approaches may augment flexibility, they do not appear to reduce the number of suppression personnel required to carry out fire-fighting operations safely.

It does seem reasonable to say that not less than two fire suppression vehicles and a command officer should respond to any structure fire and that the number of personnel responding should be sufficient to carry out the tasks indicated above, and whatever else is typically necessary in local operations, in a timely fashion. Normative data are available from some cities. Although that number is dependent on numerous factors, it is relevant to note that, in the broad spectrum of environments protected by 41 of the fire departments making up a portion of the Metropolitan Chiefs section of the International Association of Fire Chiefs, no department in the mid-1990s dispatched fewer than 13 fire fighters (including a command officer) to a reported fire in a single-family detached dwelling. The average number dispatched was 18.6, and this did not include a rapid intervention team.[1]

The 1998 National Survey on Fire Department Operations, conducted by the Research Office of the Phoenix, Arizona, Fire Department presents normative data from 335 U.S. fire departments (which together protect more than 82 million people) and 23 Canadian fire departments (which together protect more than 9 million people). In U.S. communities, detached-dwelling fire calls had 13.9 personnel on average dispatched as a first-alarm assignment; 14.6 to attached dwellings; 16.0 to commercial structures; 17.7 to schools and hospitals; 16.2 to industrial alarms; and 18.0 to highrise alarms. Canadian fire departments dispatched numbers ranging from 11.6 to 14.8 to the same types of calls. It is important to note that these averages—in most instances—do not include the members of the stand-by rapid entry (rescue) team or an incident safety officer.

Where necessitated by fire frequency or response distances, additional pumpers, ladder trucks, and tankers may be needed. Reserve apparatus is desirable not only to permit the repair of first-line equipment without reducing available fireground forces but also to provide additional fire-fighting units during major emergencies. Specialized vehicles for hazardous materials response, heavy-duty and specialized rescue—including boats—lighting equipment, breathing apparatus bottles, emergency medical response, and so on, also must be planned for. If these cannot be made available locally, then mutual aid arrangements are needed. Different types of vehicles may be necessary for the various levels of emergency medical response: first-responder or basic life support, advanced life support, and hospital transport.

Commercial, industrial, and mercantile areas generally require additional apparatus, or more, in response to the initial alarm. If properties with considerable life hazard are involved (schools, hospitals, nursing homes, etc.) additional resources should be considered for initial alarms. Especially large numbers of personnel are needed for search and rescue operations in these properties, with several fire fighters needed to "sweep and search" each floor. (See Table 4-1.)

The required fire-fighting units should arrive on scene close enough in time after the initial alarm to operate as an effective fire-fighting unit following planned tactical procedures.

Operating Personnel Safety Concerns

Over the past several years, prompted by continuing line-of-duty fire fighter deaths and injuries, additional safety protocols have been instituted by OSHA, NFPA, various state agencies, fire departments, labor organizations, and other groups. These protocols call for on-scene safety officers, incident command teams, rapid intervention

TABLE 4-1 Typical Initial Attack Response Capability Assuming Interior Attack and Operations Command Capability

---

High-hazard occupancies (schools, hospitals, nursing homes, explosives plants, refineries, high-rise buildings, and other high life hazard or large fire potential occupancies)

At least 4 pumpers, 2 ladder trucks (or combination apparatus with equivalent capabilities), 2 chief officers, and other specialized apparatus as may be needed to cope with the combustible involved; not fewer than 24 fire fighters and 2 chief officers.

Extra staffing of units first due to high-hazard occupancies is advised. One or more safety officers and a rapid intervention team(s) are also necessary.

Medium-hazard occupancies (apartments, offices, mercantile and industrial occupancies not normally requiring extensive rescue or fire-fighting forces)

At least 3 pumpers, 1 ladder truck (or combination apparatus with equivalent capabilities), 1 chief officer, and other specialized apparatus as may be needed or available; not fewer than 16 fire fighters and 1 chief officer, plus a safety officer and a rapid intervention team.

Low-hazard occupancies (one-, two- or three-family dwellings and scattered small businesses and industrial occupancies)

At least 2 pumpers, 1 ladder truck (or combination apparatus with equivalent capabilities), 1 chief officer, and other specialized apparatus as may be needed or available; not fewer than 12 fire fighters and 1 chief officer, plus a safety officer and a rapid intervention team.

Rural operations (scattered dwellings, small businesses, and farm buildings)

At least 1 pumper with a large water tank [500 gal (1.9 m^3) or more], one mobile water supply apparatus [1000 gal (3.78 m^3) or larger], and such other specialized apparatus as may be necessary to perform effective initial fire-fighting operations; at least 12 fire fighters and 1 chief officer, plus a safety officer and a rapid intervention team.

Additional alarms

At least the equivalent of that required for rural operations for second alarms; equipment as may be needed according to the type of emergency and capabilities of the fire department. This may involve the immediate use of mutual-aid companies until local forces can be supplemented with additional off-duty personnel. In some communities, single units are "special called" when needed, without always resorting to a multiple alarm. Additional units also may be needed to fill at least some empty fire stations.

---

teams for fire fighter rescue, personnel accountability systems, plus other provisions to increase responder safety. Other provisions focus on the safety of emergency medical responders, technical rescue specialists, hazardous materials responders, and so on. All of these imply that the necessary additional personnel will be present at incidents in addition to those needed to conduct typical operations.

In evaluating the adequacy of fire protection in any given area, planners must give major consideration to the ability of the fire department to handle efficiently any reasonably anticipated workload. This requires an evaluation of the possibility of simultaneous working fires and other emergencies; weather factors that may contribute to the spread of fire, the delay in response, or the possibility of slow operations at the scene; and other demographic or geographic conditions that might affect the frequency, severity, and spread of fire occurrence and the response time of initial fire-fighting units.

Where fire frequency is such that any fire company may expect two or three working fires per day, or where structures to be protected require a heavy initial response, closer geographic spacing of or increased personnel assigned to individual fire companies may be necessary. The number of other fire-fighting or related operations such as grass, brush, rubbish, and automobile fires and emergency rescue operations may also require greater-than-normal staffing of equipment and closer spacing of fire companies. Major structural fires may result when the normal first-alarm coverage in a district is depleted through coverage of these other emergencies, making remaining fire-fighting forces inadequate. Staffing fire apparatus at a level below minimum requirements can result in less effective and less safe fire-fighting performance. This factor also has an adverse effect on the number of required fire companies for various alarms, since additional fire companies must be dispatched to the scene of an emergency to provide adequate total staffing.

The desirable practice of assigning emergency medical responsibility to the fire department must be calculated into the staffing formula. It is difficult also to obtain effective teamwork and coordination with understrength crews. Some fire departments have attempted to solve this problem by supplementing their crews with part-time or volunteer fire fighters, or by providing off-duty fire fighters with tone-activated radio receivers and paying them for overtime when they respond to a fire. The on-duty personnel make the initial fire attack and holding action while off-duty personnel provide the additional assistance needed for continuing fire-fighting operations. Although useful and possibly less costly in the short run, efficiency is lost, and increased fire losses can be expected with this arrangement. Such protection should not be relied on to replace adequately the required staffing and equipment needed immediately at the scene for initial attack and rescue.

Personnel requirements are not merely a matter of numerical strength, but are also based on the establishment of a well-trained and coordinated team necessary to utilize complicated and specialized equipment under the stress of emergency conditions. Attempting to operate more fire companies than can be effectively staffed, even if some response distances must be somewhat increased, is less desirable than fewer but appropriately staffed companies. The effectiveness of pumper companies must be measured by their ability to get required hose streams into service quickly and efficiently. NFPA 1410, *Standard on Training for Initial Emergency Scene Operations*, should be used as a guide in measuring this ability. Seriously understaffed fire companies are generally limited to the use of small hose streams until additional help arrives. This action may be totally ineffective in containing even a small fire and in conducting effective rescue operations.

Consideration must be given also to maintaining an adequate concentration of additional forces to handle multiple alarms at the same fire, while still providing minimum fire protection coverage for the other areas under fire department protection. If available personnel prove adequate for routine fires but inadequate for major emergencies, arrangements should be made to supplement the fire protection coverage by calling back the off-shift personnel and by promptly calling nearby fire departments for mutual aid. Off-shift personnel may operate reserve apparatus or relieve or supplement personnel on the fireground. Fire companies not dispatched or utilized on the fire scene should be repositioned throughout the remaining area of the jurisdiction to ensure minimum response times to other alarms.

Reserve apparatus should be properly maintained and equipped, and when placed in service should be staffed to a degree commensurate with standard fire apparatus requirements. Since it may take up to 30 min or more to place reserve units in service with personnel recalled in an emergency, these reserve units should not be completely relied on to immediately provide an adequate level of fire protection services.

Concern must be shown, under legislation and local policy, for the health and safety of fire fighters and others, for environmental protection, and for the rights of those being served. Officials must demonstrate reasonable and prudent action with establishment of incident command systems, for example, and the appointment of qualified safety officers and rapid intervention team protocols.

In cases where several fire departments occupy adjacent or contiguous territories, arrangements (often termed "line response") should be made for joint response along common boundaries to high-risk hazards and for assistance in covering vacant fire stations at times of major fires. In areas where the nearest fire station or mobile unit to the incident address is not a part of the fire department district to which the address belongs, the nearest station or unit—by prearrangement—may still be dispatched to save time. This methodology is termed "closest station response." Mutual aid or mutual response should not be relied on for routine emergencies, since there could be times when local commitments will preclude the anticipated assistance. Mutual-aid agreements do not reduce the responsibility of each jurisdiction to maintain adequate facilities to handle normal fire protection needs. It also must be assumed that teamwork and tactical efficiency at a fire will be somewhat less than that expected of equal units from the same department under a united command. Often, however, specialized units (such as hazardous materials response teams) are organized to protect larger areas encompassing several fire departments.

Fairly often, consideration is given in the evaluation and planning process to the concept of merging or consolidating with one or more other departments. This unification is viewed as a possible way to attain economies of scale and possibly to increase the breadth of service delivery to the combined area. On occasion, merging has been necessary because one district has insufficient funds available to retain viability. Other times a volunteer department may not have sufficient available personnel. Although the official combining of two or more departments may be wise, careful consideration must be given to the possible gains and losses and to the understanding that the combination of two very weak departments does not typically result in a new, strong department.

Two or more departments and often a significant number of departments in a large area—such as a county or region—may benefit greatly from what is termed "functional consolidation." This concept does not have the various departments relinquishing individual autonomy but rather cooperating and, thus, achieving economies of scale and service increases. Functional consolidation ranges from joint dispatch and stations to group purchasing and training, and from regional special response teams to combined fire prevention. Intergovernmental service contracts can facilitate these cooperative ventures.

Planning for volunteer departments where there is a scarcity of response personnel may involve the addition of full-time or part-time personnel to carry part of the workload. So-called combination departments appear more and more necessary as community demographics change and workload increases. Various methodologies are used to produce response crews. These include using students and others as station "bunkers," rotating volunteer-duty shifts, using on-call personnel, and making automatic mutual-aid agreements.

In the past it was a common practice to relate the number of pumping engines and their pumping capacity, and other apparatus personnel requirements, to the population to be protected. With the industrialization of many areas and the construction of commercial shopping centers, hospitals, schools, and nursing homes in residential areas, it is possible that concentrations of life hazard and property value in areas of small or large populations may require substantial fire-fighting forces. Those with the responsibility for providing public fire protection must be prepared to cope with fire potential in any location in the jurisdiction. Fire department response requirements are now based on the water flow in gpm (L/min) that may have to be applied. A rule of thumb is to provide one company for each 250 gpm (946.25 L/min) that may be needed in an interior attack, plus personnel for rescue and other operations that need to be performed simultaneously with the advancing of hose lines.

Some may argue that it is not the public's responsibility to provide adequate fire protection to high-hazard risks that should have built-in fire protection systems. However, failure to attempt to provide fire protection for large taxable values on which the economy of a community may be based would place the community's fiscal viability at risk. Burned-out businesses may not rebuild, and then local people will lose employment. Also, fire spread to other properties is possible.

Time is another critical factor in the evaluation of public fire protection. It is generally considered that the first-arriving piece of apparatus should be at the emergency scene in 5 min of the sounding of the alarm at the fire station, since additional minutes are needed to size up the situation, deploy hose lines, initiate search and rescue, and so on. In dense urban settings, the desired response time is often shorter, and 4 min for the first-responding pumper is a rule-of-thumb maximum time for 90 percent of an urban area. An old adage says, The first 5 min of most fires is the determining factor as to whether that fire will remain a small fire or become a large fire. Although this may not always be true, delays in sounding an alarm obviously must be minimized or eliminated, as well as delays in responding and initiating rescue and attack. Time, however, cannot become the all-important factor at the expense of safety. In the interest of safety, some departments have responding vehicles run to certain calls without the use of warning devices and obeying all traffic laws ("run silent alarms").

There are numerous instances where highly specialized apparatus and equipment must be available to municipalities. One category of specialization concerns apparatus designed to handle hazardous materials, including spills of petroleum products and other chemicals that require special extinguishing agents such as foams or dry powders, and special equipment to apply these agents. These dangerous substances may be present because of airports, marinas, manufacturing or storage facilities, or transportation routes in the district. Another category of specialization includes apparatus and equipment needed because of particular structures or facilities such as research laboratories, hospitals, high-rise buildings, oil and gas wells, and seaports. In some communities, fire departments are expected to conduct specialized functions such as extricating at automobile wrecks and performing water and mountain rescue, as well as providing emergency medical services. These services also require special equipment and possibly special vehicles. The necessity for many departments to deliver at least first-responder emergency medical service, and for the delivery of a wide variety of technical rescue and disaster response services, cannot be overlooked.

As with standard fire-fighting equipment and apparatus, specialized tools cannot be used effectively and safely unless personnel are highly trained in their use under a wide variety of circumstances. Whether personnel are volunteer or career, in rural or urban areas, no plan can be implemented and no reasonable level of protection afforded to the community unless well-designed and well-managed training programs are carried out.

Progressive departments of all types and sizes also concentrate on providing a broad range of community-oriented services.

Fire Prevention

The term *fire prevention* as used here generally includes inspections, education, and equipment meant to reduce the occurrence of fire and to mitigate the effects of that fire prior to the arrival of the mobile suppression force. As an example, the installation of a sprinkler system is not designed to prevent fire but to control it in its very early stages; the "Stop, Drop, and Roll" message of the NFPA's *Learn Not to Burn*® program is clearly a mitigation rather than prevention effort. Other education efforts, such as NFPA's *Risk Watch*® program, are directed at stopping the occurrence of fire and in promoting a wide range of safe behaviors.

As noted previously, fire prevention activities are somewhat difficult to evaluate. In a real sense, if prevention activities are effective, fires and fire-related tragedies occur with less frequency. There is a reduction or absence of fire activity, and these results are statistically evident although they do not appear in dramatic news clips and photographs. Some departments do report not only dollar amounts of fire loss, but also the value of structures that were threatened by fire and thus "saved." Without careful and systematic long-term record keeping concerning the incidence of fires, fire losses, and related tragedies, the effect of prevention programs cannot be documented. Inability of fire officials to demonstrate the value of committing some additional community resources to the broad range of possible prevention activities may well result in a withdrawal of resources from prevention programs and a subsequent increase in the need for a much larger suppression budget. Rational decisions and sound recommendations concerning evaluation and planning cannot be made unless fire officials learn what changes there can be to total fire cost by reallocating resources applied to the total fire defense system.

Both evaluation and planning require recognition of the component and integrated parts of a fire prevention system. Until recent years prevention was often greatly limited or nonexistent in most smaller communities. In urban areas it was limited frequently to the periodic inspection of certain types of buildings. More modern approaches to fire prevention recognize that a comprehensive program includes all organized activities, other than suppression, that reduce the incidence of fire and fire-related losses. Ideally, these activities would be carried out in communities of every size, whether rural or urban, with appropriate adjustments made for community size, type, location, and fire history. Community- and neighborhood-based focus programs, often emanating from the local station, appear to have excellent effects.

Prevention activities may be categorized in several ways, but it is usually helpful to group them as follows:

1. Activities that relate to construction, such as building codes, the approval of building and facility plans, and occupancy certification and recertification for new occupants. Also included may be a sign-off for the presence of smoke detectors when new or old properties are sold.

2. Activities that relate to the enforcement of codes and regulations, such as inspections of certain occupancies, the licensure of certain hazardous facilities, the design of new regulations and codes, and legislation to adopt existing model codes.

3. Activities that relate to the reduction of arson, such as fire investigation, the collection of information, public education, and data related to setting fires. Included may be arson investigation and related court proceedings, and programs such as counseling for juvenile fire setters.

4. Activities that relate to the collection of data helpful in improving fire protection, such as standardized fire reporting, case histories, and fire research.

5. Activities that relate to public education and training, including fire prevention safeguards, evacuation and personal safety steps, plant protection training for in-

dustrial and other work groups, hazardous materials and devices safeguards, and encouragement to install early warning and other built-in signaling and extinguishing devices. Very popular are programs for school children, such as NFPA's *Learn Not to Burn* curriculum and self-help classes such as water safety, urban survival, and similar "Stay Alive 'Til We Arrive" projects.

An analysis of the community's fire history, conducted during the evaluation phase of the fire protection plan, will usually indicate to fire experts and citizen groups which categories need strengthening. Comparing the number of fires and fire-related incidents, plus fire loss (property, life, injury) statistics over several years as more prevention activities are phased in, provides an assessment of program effectiveness. Calculating the total cost of fire to a community (fire loss plus prevention costs plus suppression costs plus fire insurance costs) will enable the fire department and the community to estimate the efficiency or cost-effectiveness of a proposed prevention program.

PUBLIC PROTECTION CLASSIFICATIONS

Fire Department Service-Level Analysis

The public, fire and other government administrators, organized labor, fire protection organizations, and the fire service in general have for many years sought to find a generally accepted method for the evaluation of services provided to a community by its fire protection agency. The approaches have been as varied as the fire service agencies being evaluated and the parties doing the evaluation. This has resulted in the application of inconsistent methodology and criteria.

For many years material developed by NFPA has been used in the analysis of services provided by fire protection organizations. However, the application of NFPA materials has been inconsistent. The fire department accreditation program is entirely voluntary, and Insurance Services Office (ISO) reviews are typically conducted about every 10 years, unless a special request is made.

Since 1990, NFPA has continued its long history of involvement in the field of fire department evaluation through programs to provide criteria, process, and organization for the analysis of service levels provided by fire departments. With the Board of Directors' concurrence, a "Select Committee" was appointed to advise in the "Fire Department Analysis" project, with representation from the fire service, municipal management, organized labor, and the insurance industry. Appropriate recommendations to the NFPA Standards Council have been made. NFPA 1500, *Standard on Fire Department Occupational Safety and Health Program,* was modified and reissued in 1997. NFPA 1201, *Standard for Developing Fire Protection Services for the Public,* was modified and reissued in 2000. New NFPA technical committees covering fire department organization and deployment, including emergency medical services, have been formed to focus on career departments and volunteer departments.

Initial work of the Select Committee, conducted in 1990 and 1991, determined that the best program for the analysis of services would

- Consider the total scope of services provided by the fire service agency
- Set forth uniform definitions
- Be primarily performance based and user friendly
- Be founded on consensus and other standards to the greatest possible extent
- Include user input
- Be designed for specific community application

- Be capable of self-application
- Include a validation process

The Voluntary Fire Service Accreditation Program, sponsored by the IAPC, uses self-assessment and peer group review processes to assist fire departments in evaluating important aspects of their organization and its operations.

A group of experienced fire service personnel, meeting in October 1996 at the Wingspread IV Conference, listed the following emerging issues of national importance to the fire service: customer service, managed care, competition and marketing, service delivery standards, fire fighter wellness, and labor-management relations.

The following were listed as earlier and ongoing issues of national importance to the fire service: leadership in changed environments; expanded prevention and public education efforts; professionalized training and education; increased use of fire and life safety detection, alarm, and extinguishment systems; the formation of strategic partnerships; the collection of relevant data; and the protection of the environment.

Insurance Services Office (ISO) Fire Suppression Rating Schedule (FSRS)

Although not all states in the United States use the existing Insurance Services Office (ISO) grading schedule, and it is not used in other nations, it is applied to many departments in most states approximately every 10 years. The purpose is to aid in the calculation of fire insurance rates and is not for property loss prevention or life safety purposes.

The service focus of ISO has broadened considerably over the past few years. The ISO prepares Public Protection Classification (PPC) reports for nearly 43,000 fire protection districts in the United States. In addition ISO has created a Building Code Effectiveness Grading Schedule to determine how well a municipality enforces its building code.

Furthermore, the new ISO Community Outreach Program collects information on essential fire protection features within a community. This information will lead to sound benchmarking data, permitting individual communities to better assess their own safety level and to respond accurately when completing the Commission on Fire Accreditation International's self-assessment forms.

In a January 2001 survey of 502 U.S. fire chiefs and fire department officials, 92 percent responded that their ISO Public Protection Classification number is a direct reflection of the improvements made in their community. They reported that important local uses of the ISO PPC program are

- Helping save lives and property (90%)
- Helping save money on fire insurance (67%)
- Planning for and budgeting for changes in community fire protection (61%)

The older form of the grading schedule (1974) contains more categories and more items than the schedule in current use. However, the older form is still applied in one or more states and is still used by some community officials as a reference tool in self-evaluations.

In Section I of the ISO grading schedule, the result and classification apply to properties with a needed fire waterflow of 3500 gpm (134,248 L/min) or less. Private and public properties with larger needed flows are individually evaluated in Section II.

With improved ratings, fire insurance companies that subscribe to the ISO rating system may lower commercial fire insurance premiums and may lower residential rates in certain instances. At least one insurance company is using a somewhat broader approach to determine premium costs, by reviewing total insurance costs for all hazards across zip code or other defined areas.

The *Grading Schedule for Municipal Fire Protection,* developed originally by the National Board of Fire Underwriters (NBFU) and continued by its successor, the American Insurance Association, and then by the ISO, has provided a guideline for municipalities to classify their fire defenses and physical conditions. The gradings obtained under the schedule are used in establishing base rates for fire insurance purposes. The schedule has been subject to change with the state of the art, and sweeping changes were made in the 1980 edition with the development of a revised FSRS, and with additional changes in 1995 and 1998. Other changes will continue to be made as warranted. Under certain circumstances, credit may now be given for property within five road miles of a fire station, even though a municipal water supply system is lacking.

The current ISO grading schedule reviews and correlates those features of public fire protection that have a significant effect on minimizing fire damage. Credit is given for existing fire protection, instead of debit for what is not in place.

The Fire Suppression Rating Schedule (FSRS) produces 10 different Public Protection Classifications, with Class 1 receiving the most rate recognition and Class 10 receiving no recognition. The FSRS simply defines different levels of public fire suppression capabilities that are credited in the individual property fire insurance rate relativities.

Starting in 1975, on a state-by-state basis following Insurance Department approval, ISO implemented the Commercial Fire Rating Schedule (CFRS), which was a major revision to the method used to develop individual property rate relativities. The CFRS reviews and correlates the construction, occupancy, exposures, and private and public fire protection (represented by the Public Protection Classification number). This correlation allows development of an equitable rate relativity applicable to the individual property. To this rate relativity, statistical experience adjustments are applied either by ISO or by affiliated companies to produce the applicable fire insurance rate. The FSRS represents a revision in the method used to derive the Public Protection Classification number used in the CFRS. The Public Protection Classification number is also used as a rate relativity variable for most class-related properties, in addition to construction and occupancy variables.

The *Grading Schedule for Municipal Fire Protection,*[2] although a much-improved system from previous editions, was not developed as an integral part of the individual property rating system. The previous schedules were somewhat independent primarily due to their historical development by the NBFU, which was not an insurance rating organization. The previous schedules were used more to quantify underwriting information but did define different levels of public fire protection that could be used for a specific rating.

The FSRS is designed to assist in an objective review of those features of available public fire protection that have a significant influence on minimizing damage once a fire has occurred. This revision ties logically to the review of contributive and causative hazards that can be performed with the CFRS.

FSRS Class Groupings

As stated earlier, the ISO prepares Public Protection Classification (PPC) reports for nearly 43,000 fire protection districts in the United States. Districts include political jurisdictions identified as counties, cities, towns, villages, municipalities, and fire districts. A district may support an organized fire department or contract with an existing fire department for fire suppression services. Each PPC report is prepared using an information base established from doing a city grading evaluation conducted by one of 160 regional ISO field representatives, who are highly trained in the application of the

FSRS. ISO conducts municipal surveys in 45 of the 50 states. Mississippi and Washington state have elected to use the 1974 edition of the FSRS. Hawaii, Idaho, and Louisiana administer the current edition of the FSRS through state rating organizations. The District of Columbia and New York City are not graded using the FSRS.

The scope, objectives, and methods of application for the current FSRS are significantly different from previous grading schedule documents. Technological change in the real world is reflected in all elements of the current grading schedule pertaining to the three major items of coverage that follow:

- Receiving and handling fire alarms
- Fire department
- Water supply

Today, the FSRS provides an objective tool for the review of city public fire suppression facilities, equipment, and programs. Credits are assigned for specific fire protection features covered by the grading schedule. Calculating deficiencies on individual items has been dropped; the percentage of adequacy is now determined for each item. Each grading schedule survey involves an extensive review of fire department records, field surveys of the city, and the testing of fire protection equipment and municipal water-supply systems.

Information gathered during the field survey is applied to the individual items in the grading schedule using quantitative analysis. The city's fire suppression potential capability is then assigned a Public Protection Classification number according to the following class groupings:

- Classes 1–8 are the protected property classification. All properties assigned a Class 1 through 8 have a recognized fire department with engine company response limited to 5 travel miles (8 km) in most states and a recognized water delivery system. The minimum water supply from each credited fire hydrant is 250 gpm (946 L/min) for a 2-hr duration; therefore, a recognized water system must also deliver the same minimum flow and duration.

- Class 9 is the semiprotected property classification. All properties assigned a Class 9 have a recognized fire department generally limited to 5 travel miles (8 km). However, structural property is beyond 1000 feet (305 m) of a recognized water supply. The current FSRS provides a documented method for fire departments to deliver adequate water to fire sites using mobile water tankers to permit structural property in Class 9 areas to qualify for a protected property classification (Classes 1 through 8). Water delivery demonstration projects have improved Class 9 PPC districts all the way down to a Class 4.

- Class 10 is the unprotected property classification for structural property. Class 10 property is generally located beyond 5 travel miles of a recognized fire station, regardless of available water supply. Only a selected few insurers will write insurance policies for Class 10 property. When available, the insurance rates are very high compared to other property classes.

Scope and Content of the FSRS

The grading schedule is divided into major sections: the public fire suppression and the individual property fire suppression.

Section 1: Public Fire Suppression.   This section is applied to develop a Public Protection Classification (PPC) for all class-rated properties and for specifically rated properties in a city with a needed fire flow (NFF$_i$) of 3500 gpm (13,249 L/min) or less. The section is, in turn, divided into three major items for evaluation:

1. *Section 400: Receiving and Handling Fire Alarms.* Ten percent of an overall city's grading is based on how well the alarms are received and the fire department is dispatched. The assigned ISO field representative evaluates the alarm-dispatch center, looking at the telephone service capability, the number of telephone lines coming into the center, the listing of emergency numbers in the area telephone book, and the number of operators on duty in the center at all times. The ISO review also examines all dispatch circuits and the electronic methods used to notify fire fighters of the location of fire incidents. The ISO uses NFPA 1221, *Standard for the Installation, Maintenance, and Use of Emergency Services Communications Systems,* as a guide for grading public sector fire alarm systems.

2. *Section 500: Fire Department.* Fifty percent of the city's overall grading is based on the fire department evaluation. The grading schedule considers a first-alarm assignment to be a minimum of two engine companies and one ladder-service company to all structure fires. ISO evaluates the distribution of engine companies and ladder-service companies for response areas of the city in accordance with whether the built-upon area of the city has a first-due engine company within 1.5 miles (2.4 km) and a ladder-service company within 2.5 miles (4 km).

   ISO also checks to determine that the permanently mounted pumps on the fire apparatus are tested regularly and inventories are taken of each engine company's complement of fire hose, nozzles, self-contained breathing apparatus (SCBA), and small equipment items. Furthermore, ISO checks on the number and type of ladders, including both ground and aerial ladders, plus service equipment that includes salvage covers, power saws, ventilation equipment, and lighting equipment. Finally, ISO reviews fire company records to determine

   - Classification and extent of training provided to fire company personnel
   - Actual personnel who participate in training programs
   - Number of fire fighters who respond to structure fires
   - Level of building familiarization and documented prefire planning conducted by fire personnel

   ISO references several NFPA standards in the evaluation process of fire departments.

3. *Section 600: Water Supply.* Forty percent of the grading is based on the city's water supply. ISO examines the following three components of each water system to assure that sufficient water capacity, flow rate in gallons per minute and pressure at 20 psi residual pressure, is available at selected sites throughout the city.

   - Water-supply works
   - Water-supply mains feeding fire hydrants
   - Fire hydrant installation, maintenance, and inspection

   The water-supply works and pipe distribution system accounts for 35 percent of the entire city grading. To accomplish an analysis of a given water system, the ISO field representative:

   - Examines whether sufficient water is available for fire suppression beyond the city's maximum daily consumption.
   - Surveys all components of the water supply system, including stationary pumps, filtration capacity to provide potable water, and potable water storage to supply water mains.

- Observes fire-flow tests at representative locations in the city to determine the rate of flow provided by water mains.

- Counts the distribution of fire hydrants up to 1000 feet (305 m) from representative categories of property throughout the city. The classes of property evaluated include but are not limited to industrial, commercial, educational, religious, health care, and residential.

- Considers the size, type, and installation of fire hydrants, along with the operating condition of all fire hydrants.

Fire hydrants should receive semiannual inspection, as outlined in the American Water Works Association Manual 17, *Installation, Field Testing, and Maintenance of Fire Hydrants*. This analysis is worth 5 percent of the entire grading.

Section II: Individual Property Fire Suppression.   This section develops Public Protection Classification for specifically rated properties that have a needed fire flow between 4000 gpm (15,142 L/min) and 12,000 gpm (45,425 L/min). The following are supporting topics on Public Protection Classification.

1. *Preparing for an ISO Grading Evaluation.* A city can maximize earned credits through a systematic and proper preparation for an ISO grading evaluation. Doing so involves having in place current maps, inventories, equipment test records, and personnel reports for the ISO field representative to review and evaluate in accordance with each item documented in the FSRS. The chief executive officer (e.g., mayor, city manager) of a city can request from ISO a *Public Protection Classification-Evaluation Resource Manual* that details the information requested when an ISO field representative visits a city.

2. *Impact of a City's Public Protection Class on Fire Insurance Premiums.* Theoretically, the better a city's classification, with Class 1 being the best class, the lower will be both insurance rates and insurance premiums when compared to a higher class number. This is generally true for a commercial property that is specifically rated by the insurance industry. Other factors, however, enter into the premium calculation, including the following:

   - Fire protection equipment such as installed fire extinguishers, early warning detection and fire alarm systems, smoke control systems in some occupancies, and, most importantly, the installation of automatic sprinkler systems

   - Fire loss, or loss costs, to the insurance industry in the city or county where the building risk is located

Furthermore, some constant-risk commercial property, such as a drugstore, and all one- and two-family dwellings are grouped by insurers into PPC sets as Classes 1 to 6, Classes 7 and 8, Class 9, and Class 10. The loss costs in each set are so similar that individual class rate structures cannot be justified on the basis of underwriting experience. Under the preceding criteria, a residential home owner would not receive a premium reduction if a given city improved to a Class 5 or better. However, the general percentage reduction that commercial property owners can expect in premium reduction percentages is shown in Table 4-2. Finally, it needs to be recognized that individual insurance companies may file with state insurance commissions for "rate deviations" from "standard rates" for specific classes of property based on that company's loss experience and insurance reserves. This underscores the highly competitive nature of the property insurance industry today.

TABLE 4-2   Percentage Reductions for
Commercial Property Insurance

City Class Change	Percent of Premium Decrease
Class 10 to Class 9	15
Class 9 to Class 8	9
Class 8 to Class 7	5
Class 7 to Class 6	5
Class 6 to Class 5	5
Class 5 to Class 4	5
Class 4 to Class 3	8
Class 3 to Class 2	3
Class 2 to Class 1	2

PLANNING

Whenever a community—rural, suburban, or urban—considers its fire defenses, it must scrutinize the past and present and make predictions or forecasts for the future. Reviewing the past is called *data analysis* and depends on good record keeping. *Evaluation,* which is looking at the present, requires the ability to examine a situation objectively. The process of forecasting future conditions and preparing for them requires that a planning process be followed. This *planning* process results in a plan and its implementation, so that future challenges to the community are met. As the plan is implemented, the process must include the establishment of a *feedback loop,* providing a continuing assessment of how well the plan is contributing to successful completion of goals and objectives, and feeding revised data back into the plan so continuing redesign occurs.

Fire protection organizations, and especially fire departments, need to develop several kinds of plans related to fire prevention and fire suppression. These plans should be quite specific, directed at clearly defined goals, and operational over a relatively brief time period (usually from 1 to 5 yr). Typically, such plans are internal to the department and do not involve broad-based planning groups from the outside. Examples of these types of plans, which are most often technical in nature, are apparatus replacement plans, training program plans, revised initial-response plans, plans for a special hazardous-materials attack unit, and plans for adapting fireground procedures to incorporate the use of larger-diameter hose. However, once department planning begins to consider aspects of fire protection that will have an impact on external groups, those groups will need to be consulted and incorporated into the planning process. Fire department planning, for example, must consider land-use planning (zoning), water department planning, building code enforcement, and so on.

Examples of fire department planning that require the early involvement of other groups are station relocations or closings, building inspection programs, public education programs, and changes in the scheduling of work platoons. These plans, although involving some other groups, are still fairly narrow in scope and usually can be formulated over a relatively short time.

A third type of planning, often called comprehensive, master, or strategic planning, addresses the total community fire protection problem, incorporating both prevention and suppression, and obviously involves many community agencies and organizations, perhaps even county, state, and federal agencies. Comprehensive planning is a necessity for communities and is aimed at integrating all community efforts at prevention and suppression, and improving efficiency and cost-effectiveness of those activities. Improved total community fire protection is the goal of this planning. Its degree of success must be measured by figures relating to the total cost of fire to the community, and not just in gains for one subsystem. Comprehensive plans often consist of a number of subplans from various agencies that are developed at the same time as part of a larger, total process, and that fit together to make a comprehensive and integrated plan.

Comprehensive plans have clearly stated goals with agreed-upon ways of measuring their attainment. These overall goals are reached through overall strategies acceptable to all involved agencies and to the citizens who must pay for the fire protection system. Each goal is composed of some number of subgoals or objectives, and for each objective there is a tactic designed to reach that objective. All objectives lead to the accomplishment of the overall strategies. When the objectives and tactics are laid out on a timeline, the overall time required to implement the comprehensive plan is then known, and the timing for attaining each objective is apparent.

Fire protection has been largely a local responsibility, and for good reasons it seems destined to remain so.* Each community has a set of conditions unique to itself. To be adequate, the fire protection system must respond to local conditions, and especially to changing conditions. Planning is the key: Without local-level planning, the fire protection system is apt to be ill-suited to local needs and unadaptable to the changing needs of the community.

Excellent fire protection (for example, in the form of automatic extinguishing systems such as residential sprinklers or in the form of technically advanced and trained fire-fighting forces) is technically available and certainly can be provided with the resources of most communities. Even with considerable public support, however, this protection may require several years to attain. In the meantime, in every fire jurisdiction (whether a municipality, county, or region), fire protection goals must be set and plans made to achieve those goals. The issues below discuss some of the concepts to be defined in setting these standards.

Goal-Setting Concepts

Adequate Level of Fire Protection.   The question of "adequacy" is addressed not only in day-to-day needs but also in major contingencies that can be anticipated for future needs as well. A definition of "optimal" protection is needed—in contrast to "minimal" protection, which fails to meet contingencies and future needs, and "maximal" protection, which is usually more expensive than a community can afford.

Comprehensive planning must include contingencies drawn from an analysis of community hazards. This process of hazard identification and analysis is crucial to fire department planning.

Reasonable Community Costs.   Fire, both as threat and reality, has its costs, including deaths, injuries, property losses, hospital bills, and lost tax revenues, plus the costs of maintaining fire departments, paying fire insurance premiums, and providing built-in fire protection. Each community must decide on an appropriate level of investment in fire protection. Some costs that are beyond the public's willingness to bear may be transferred to the private sector (as when buildings over a certain size or height or with a certain occupancy are required to have automatic extinguishing systems).

---

*Some of the information that follows in this chapter has been extracted in whole and in part from *America Burning*[3] and *America Burning Revisited*.[4]

Acceptable Risk.   A certain level of fire loss must be accepted as tolerable simply because of limited resources of a community. Conditions that endanger the safety of citizens and fire fighters beyond the acceptable risk must be identified as targets for mitigation.

Consideration of these matters helps to determine what functions and emphasis should be assigned to the fire department, other municipal departments, and the private sector both now and in the future. It helps to define new policies, laws, or regulations that may be needed. Most importantly, consideration of these matters makes it clear that fire safety is a responsibility shared by the public and private sectors. Because the fire department cannot prevent all fire losses, formal obligations to have built-in fire protection fall on owners of certain kinds of buildings. For the same reason, private citizens have an obligation to exercise prudence with regard to fire in their daily lives. But prudence also requires education in fire safety, and the obligation to provide that education appropriately falls in the public sector, chiefly the fire department. The public sector (again, chiefly the fire and building code enforcement departments) also has an obligation to see that requirements for built-in protection in the private sector are being met. A fire department, then, has more than one responsibility—and the aforementioned responsibilities are not exhaustive.

Functions of Fire Protection Agencies

The following are significant functions for which fire protection agencies typically have primary or significant roles.

Fire Suppression.   Fire fighters need proper training and adequate equipment to save lives, extinguish fires quickly, and to ensure their own safety.

Specialized Emergency and Disaster Services.   These include hazardous-materials incidents, floods, earthquakes, multiple-vehicle accidents, cave-ins, collapsed buildings, volcanic eruptions, searches for lost persons, attempted suicides, a variety of specialized technical rescue services, and so on.

Emergency Medical Services.   Capabilities needed during fires and other emergencies include first aid, resuscitation, and possibly advanced life support (paramedical services). (The term *paramedical services* means emergency treatment beyond ordinary first aid, performed by fire service personnel under supervision—through radio communication and preapproved treatment protocols, for example—of a physician.)

Fire Prevention.   This includes approval of building plans and actual construction; inspection of buildings, their contents, and their fire protection equipment; and investigation of fire cause and spread to guide future fire prevention priorities and determine when the crime of arson has been committed.

Fire Safety and Dangerous-Situation Education.   Fire departments have an obligation to bring fire safety and dangerous-situation education not only into schools and private homes but also into occupancies such as restaurants, hotels, hospitals, and nursing homes that have a greater-than-average fire potential or life safety hazard. These programs may include such topics as swimming pool safety, babysitter training, latchkey child safety, and so on.

Deteriorated Building Hazards.   In coordination with other municipal departments, fire departments can work to abate serious hazards to health and safety caused by deteriorated structures or abandoned buildings.

Regional Coordination.   Major emergencies can exceed the capabilities of a single fire department, and neighboring fire jurisdictions should have detailed plans for coping with such emergencies. But effectiveness may also be improved through sharing of day-to-day operations—such as, for example, an areawide communication and dispatch network or a joint training facility.

Data Development.   Knowledge of fire department performance and how practices should change to improve performance depends on adequate record keeping. Computers play a key role in fire protection and emergency management planning.

Community Relations.   Fire departments are representative of the local community that supports them. The impression they make on citizens affects how citizens view their government. Volunteer departments dependent on private donations must, of course, also be concerned with community relations. Moreover, since fire stations are strategically located throughout the community, they can serve as referral or dispensing agencies for a wide range of municipal services.

As communities set out to improve their fire protection, they must not consider the fire department alone. The police have a role in reporting fires and in handling traffic and crowds during fires. The cooperation of the building department is needed to enforce the fire safety provisions of building codes. The work of the water department in maintaining the water system is vital to fire suppression. In fire safety education, the public schools, the department of recreation, and the public library can augment the work of the fire department. Future community development and planning will influence the location of new fire stations and how they will be equipped.

The foregoing nine functions are just the obvious examples of interdependence. Although it may seem trivial, the manner in which house numbers are assigned and posted, for example, can affect the ability of a fire department to respond quickly and effectively to emergencies, as do many other seemingly unrelated topics.

Master Planning

Fire protection is only one of many community services. Not only must it compete for dollars with other municipal needs such as the education system and the police department, but in planning for future growth the fire protection system must account for the changes in progress elsewhere in the community. For example, if a deteriorated area is to be torn down and replaced with high-rise apartment buildings, the fire protection needs of that area will change. Changes in zoning maps will also change the fire protection needs in different parts of the community.

To cope with future growth, local administrators are turning increasingly to the concept of master (comprehensive) planning of municipal functions. Such plans include an examination of existing programs, projection of future needs of the community, and a determination of methods to fill those needs. They seek the most cost-effective allocations of resources to help ensure that the needs will be met.

A major section of a community's general plan of land use should be a master plan for fire protection, which should be written chiefly by fire department managers. This plan should, first of all, be consistent with and reinforce the goals of a city's overall general plan and its time frame. For example, managers should plan the deployment of personnel and equipment according to the kind of growth and the specific areas of growth that the community foresees. It is critical that the comprehensive plan determine and set goals and objectives for the fire protection system and the fire department in terms that are understandable to the citizens being protected.

Having established goals, the department officers should use the plan to establish some form of management systematic and quality assurance within the fire department. Management is most effective when each person is aware of how tasks fit into the overall goals and is committed to getting specific jobs done in a specified time.

Because fire departments exist in a real world where a variety of purposes must be served with a limited amount of money, it is important that every dollar be invested for maximum return on investment. The fire protection master plan should not only seek to provide the maximum cost-benefit ratio for fire protection expenditures, but it should also establish a framework for measuring the effectiveness of these expenditures. Lastly, the plan should clarify the fire protection responsibility for other groups, both governmental and private, in the community.

Often, as the result of a comprehensive planning effort, a community will formulate a "developmental plan." This document provides a schedule for implementing those changes in the fire protection system deemed appropriate by the community.

Key questions for developmental plans include:

1. What are the fire protection, rescue, emergency medical, disaster response, and safety education needs?
2. What organizational structure and what resources are currently available to meet those needs?
3. Is there a disparity between what is available and what is needed?
4. What will the community profile be like in 5 to 10 yr?
5. What will be needed then to provide adequate protection?
6. What is and what will be the financial resource base?
7. What options are and will be available to enhance protection, or to keep it at an adequate level?
8. How can changes be phased in to gain community goals and cost-effectiveness?

Devising a Fire Protection Plan

Key questions to be asked by those planning for fire protection are:

1. Why is planning necessary for us at this time?
2. What do we need to start the process, and are the necessary groups committed to the process?
3. What are the necessary steps in the planning process?
4. How will the plan be implemented?
5. Are all aspects of the plan legally possible and enforceable?
6. How will the plan be evaluated? Will it be a part of the integrated emergency management plan of the community?
7. How will feedback be gathered and the plan modified and updated?

In *Introductory Summary: Fire Prevention and Control Master Planning,* the U.S. Fire Administration[5] points out the following, in providing its overview of the planning process:

Master planning is a participative process which should result in the establishment of a fire prevention and control system which is goal-oriented, long-term, comprehensive, provides known cost/loss performance, and adapts continually to the changing needs of your community.

Master planning should consider all community elements . . . related to fire prevention and control system elements.

Master planning involves the participation of all parties interested in the development of a defined cost/loss relationship. . . .

Master planning allows you . . . to systematically analyze fire prevention and control through common-sense procedures. . . . Master planning has three phases: preplanning, planning, and implementation. The preplanning phase gets necessary commitments, committees, estimates and schedules, and go-ahead approvals. The planning phase gathers and analyzes data, sets goals and objectives, determines an acceptable level of fire protection service, identifies alternatives, and constructs the plan. The implementation phase never ends, because the plan is ongoing and always being revised and updated.

The following can serve as guidelines to fire department administrators for developing and presenting a master fire protection plan as part of the comprehensive master plan outlined earlier.

Phase I

1. Identify the fire protection problems of the jurisdiction.
2. Identify the best combination of public resources and built-in protection required to manage the fire problem, within acceptable limits:
   a. Specify current capabilities of and future needs for public resources.
   b. Specify current capabilities and future requirements for built-in protection.
3. Develop alternative methods that will result in trade-offs between benefits and risks.
4. Establish a system of goals, programs, and cost estimates to implement the plan:
   a. Develop department goals and programs, including maximum possible participation of fire department personnel of all ranks.
   b. Provide goals and objectives for all divisions, supportive of the overall goals of the department.
   c. Strive to develop management development programs that increase acceptance of authority and responsibility by all fire officers as they strive to accomplish established objectives and programs.

Phase II

1. Develop a definition of the roles of other government agencies in the fire protection process.
2. Present the proposed municipal fire protection system to the city administration for review.
3. Present the proposed system for adoption as the fire protection element of the jurisdiction's general plan. The standard process for development of a general plan provides the fire department administrator an opportunity to inform the community leaders of the fire protection goals and system and to obtain their support.

Phase III

In considering the fire protection element of the general plan, the governing body of the jurisdiction will have to pay special attention to:

1. Short- and long-range goals
2. Long-range staffing and capital improvement plans
3. Code revisions required to provide fire loss management

Phase IV

The fire loss management system must be reviewed and updated as budget allocations, capital improvement plans, and code revisions occur. Continuing review of results should concentrate on these areas:

1. Did fires remain within estimated limits?
2. Should limits be changed?
3. Did losses prove to be acceptable?
4. Could resources be decreased, or should they be increased?

SUMMARY

Public fire protection should consist of a broad range of safeguards, programs, and activities ranging from plan review and code enforcement to provisions for fire suppression and public safety education. Careful consideration of local hazards and demographics, plus analysis of data related to protection and suppression are necessary for the design and maintenance of an effective, cost-effective system.

Local conditions, regulatory orders, and national standards dictate the type and level of prevention and suppression/rescue provisions necessary and appropriate for a community. The time required for response and the number and types of emergency responders and vehicles should match local needs and conform to legal and industry standard requirements. Judgments about the adequacy of local prevention and emergency response provisions can be made by local authorities, Insurance Services Office (ISO) representatives, fire department accreditation teams, or other experts. Fire insurance premiums for residential and commercial property most often reflect periodic assessments of local fire protection provisions.

Municipal planning for fire protection, fire department–based emergency medical services, hazardous materials incident response, and technical rescue services requires a review of past and present data, plus knowledge-based predictions of community change and development. Agreements concerning the level of local emergency service required then lead to decisions concerning the type, size, and deployment of suppression/rescue forces.

REFERENCES

1. National Commission on Fire Prevention and Control, *America Burning: The Report of the National Commission on Fire Prevention and Control*, U.S. Government Printing Office, Washington, DC, 1974.
2. Insurance Services Office, *Grading Schedule for Municipal Fire Protection*, Insurance Service Office, New York, 1974, and FSRS 1998.
3. NCFPC, *America Burning*, U.S. Fire Administration, Washington, DC, 1974.
4. United States Fire Academy, *America Burning Revisited*, U.S. Fire Administration, Washington, DC, 1987.
5. National Fire Safety and Research Office, *Introductory Summary: Fire Prevention and Control Master Planning*, U.S. Department of Commerce, National Fire Safety and Research Office, Washington, DC.

5

Risk Management Planning

Jonathan D. Kipp

The use of a risk management plan or process, both for emergency incident scene operations and other fire department activities, is an important tool in an occupational safety and health program. Planning can include city or town risk management personnel, insurance carriers, and other city or town departments. If the fire department participates in a city/town or county or state plan, it must ensure that a separate plan is developed and used for emergency incident operations. This chapter outlines the steps in the development of a risk management plan and includes a sample risk management plan provided by the Virginia Beach Fire Department.

INTRODUCTION

Effective risk management can save your life. Ineffective risk management can cost you your life. These principles are true regardless of where you are or what you are doing.

It is impossible to live our lives in the absence of risk. Even if we never leave home and have no contact with the outside world, there are risks: starvation, a plane crashing on our house, falling and breaking a leg. When we do leave home, the risks are far more numerous and potentially dangerous: car crashes, violence, disease. The list is endless. It is how we manage these risks that determines our success, survival, and enjoyment of life.

Risk and the management of it are not foreign to us. In fact, they are such an integral part of our lives that we may never have consciously considered them. We've all heard, and probably used, terminology such as "it's not worth the risk," and "risky business." However, do we know what we actually mean when we say those words?

Many people believe that risk management is an administrative exercise. It is not. Rather, it is a decision-making process that each and every one of us uses continually in our everyday lives. It provides a systematic method for making choices. For example, before we step into the shower, most of us check the water temperature. Why?

Jonathan D. Kipp, CSP, is the loss prevention manager of Primex[3]. He is a principal member of the NFPA Technical Committee on Fire Service Occupational Safety and the NFPA Technical Committee on Fire Service Administrative Risk Management. He was also a contributing author to the first edition of the NFPA 1500 handbook and co-authored *Emergency Incident Risk Management: A Safety and Health Perspective* with Battalion Chief Murrey Loflin of the Virginia Beach Fire Department.

Source: Originally appeared as Supplement 7, "Risk Management Planning," in Stephen N. Foley, ed., *Fire Department Occupational Health and Safety Standards Handbook*, NFPA, Quincy, MA, 1998.

We don't want to risk being burned. We effectively manage that risk by checking the water temperature first. The key is to better understand the process that led us to check the temperature and then to more deliberately apply it in our work lives as well.

NFPA 1500, *Standard on Fire Department Occupational Safety and Health Program*, first included language pertaining to risk management in the 1992 edition. The committee members believed that emergency service organizations (ESOs) needed guidance on the development of a comprehensive plan under which a safety program would logically fit. Since risk management as a discipline has been used successfully for years by other organizations, it seemed the appropriate vehicle for fire departments as well. In the 1997 edition of NFPA 1500 risk management is included in two chapters.

The standard includes language that requires the organization to develop and implement a plan for effectively managing its risks. Such a plan would be the umbrella under which the various other components of NFPA 1500 would fit. The section that addresses risk management for emergency operations is found in Chapter 6 of NFPA 1500. In a nutshell, the requirements of NFPA 1500:

- Stress the practicality of applying effective risks management techniques
- Define risk and the risk management process
- Highlight the risk management differences between an organization that responds to emergencies and one that doesn't
- Outline the different types of risk management that an emergency services organization can utilize
- Address risk management for emergency incidents, including the components of the risk management toolbox.

The Risk Management Process

The effective management of risk is a process, not a static event or a document that resides in a binder. As such, it has various components and definitions associated with it. While the definitions vary, their meaning does not. A hazard is something that increases the chances of an accident occurring. Without a hazard, there is no risk. For example, in the assessment of risk in taking a shower, the obvious risk to be considered is the possibility of being burned by hot water (the hazard). Or, to put it another way, hot water can be hazardous to us. Without the hazard (hot water), there is no risk (of being burned).

Following the same logic, we use a risk management process to address our risk to the hazard. The process is composed of a series of logical steps, similar to a decision tree, that lead us to a course of action that, hopefully, will be as risk-free as possible.

Tolerance for Risk, or Risk Management in Reality

It is important to put risk management into the appropriate context. From the perspective of the organization (fire department), risks are numerous and varied. The organization, under the guidance of its leaders, attempts to put into place appropriate programs, policies, and procedures that will help to protect its resources (people, facilities, apparatus, equipment, etc.). However, each program, policy, and procedure depends on effective execution by an individual—and everybody is different.

On a personal level, each individual has a different tolerance for risk. Some of us will take more risks than others. That is why some people choose skydiving as a hobby, while others are more comfortable collecting stamps. Apply that premise to a group of fire fighters, and the same will be true: Some of the group will be more comfortable in a hazardous, or risky, situation than will others. The organization's dilemma is to establish policies and procedures that will apply to everyone when each of us is different. There is no "risk tolerance" test that we can administer to definitively measure somebody's capabilities.

Therefore, it is vitally important that all fire fighters be trained as effectively as possible. Training, in conjunction with the appropriate personal protective equipment, policies, and procedures, will help to direct each fire fighter. Beyond those factors, however, others such as physical condition, judgment, and, ultimately, the individual's tolerance for risk at that very moment will determine what actions are taken.

An additional burden is time, or the lack of it. Decisions that will determine which actions are ultimately taken must be made quickly. Lives depend on it, and there is usually no second chance. Consider, for example, an off-duty fire fighter who is faced with a civilian in trouble—a rafter thrown from a raft and being carried away in a fast-running river. Immediate action is required. What does the fire fighter do? By training and experience, she knows to instantaneously size up the risks involved for both the rescuer and the victim (drowning, hypothermia, trauma for hitting rocks), measures them against the potential benefit (saving the civilian), calculates the odds of success (likelihood of reaching the victim, likelihood of the victim surviving), and decides what to do. Her choices can range from going to the nearest phone and dailing 911 (if she has concluded that the odds are not in her or the victim's favor and the risks greatly outweigh the benefits), to diving in and attempting to make the rescue (if she has concluded that the odds are in her favor and that she has a reasonable chance of reaching the victim, is a strong enough swimmer, has the appropriate rescue skills, and, ultimately, feels up to it at that moment). There is no policy or guideline that can dictate which action should be chosen. There are just too many variables, many of them human factors that can only be calculated in the moment.

SUMMARY STEPS TO DEVELOPING A RISK MANAGEMENT PLAN

Step One

(a) Conduct a brief overview of the program concept with the fire department's administrative staff
(b) Review department's annual accident and injury data
(c) Randomly interview staff and personnel
(d) Formulate written report

Step Two

(a) Understand the risk management process

Step Three

(a) Understand pre-emergency risk management

1. Policy and procedures
2. Training and education
3. Response
4. Personnel protective equipment
5. Incident management system
6. Health and safety officer

(b) Understand risk management plan

    1. Identification

    2. Frequency and severity

    3. Priority

    4. Control measures

    5. Monitoring

Step Four

Provide written conclusions and recommendations to the fire chief, administrative staff, and Occupational Safety and Health Committee

SAMPLE PLAN: VIRGINIA BEACH FIRE DEPARTMENT RISK MANAGEMENT PLAN

Purpose

The Virginia Beach Fire Department has developed, implemented, and operates a risk management plan. The intent of this plan is as follows:

(a) To effectively serve our customers both internally and externally

(b) To reduce the severity of occupational risks encountered by our members that could have harmful consequences to service delivery

Scope

The Risk Management Plan is intended to comply with the requirements of NFPA 1500, *Standard for a Fire Department Occupational Safety and Health Program*, specifically with the following paragraphs:

- **2-2.1** The fire department shall adopt an official written risk management plan that addresses all fire department policies and procedures.
- **2-2.2** The risk management plan shall cover administration, facilities, training, vehicle operations, protective clothing and equipment, operations at emergency incidents, operations at nonemergency incidents, and other related activities.
- **2-2.3** The risk management plan shall include at least the following components:
    - (a) *Risk identification.* Potential problems
    - (b) *Risk evaluation.* Likelihood of occurrence of a given problem and severity of its consequences
    - (c) *Prioritization of risks.* Prioritize risks based upon analysis factors
    - (d) *Risk control techniques.* Solutions for elimination or mitigation of potential problems; implementation of best solution
    - (e) *Risk management monitoring.* Evaluation of effectiveness of risk control techniques

Methodology

This plan employs a variety of strategies in order to meet the variety of objectives. The specific strategies are identified as follows:

(a) Records and reports on the frequency and severity of accidents, injuries, and occupational illnesses

(b) Reports received from the city's insurance carriers and workers' compensation

(c) Specific occurrences that identify the need for risk management

(d) National trends and reports that are applicable to the department

(e) Knowledge of the inherent risks that are encountered by fire departments and specific situations that are identified in Virginia Beach

(f) Additional areas identified by department members

Plan Organization

The plan organization utilizes the following tactics to strive toward compliance with the risk management plan:

(a) Identification of the risks that members encounter or may be expected to confront

(b) Identification of nonemergency risks including such functions as training, physical fitness, returning from an emergency incident, routine highway driving, station activities (vehicle maintenance, station maintenance, daily office functions, etc.)

(c) Identification of emergency risks including such fireground operations, EMS, hazardous materials incidents, and special operations; also including the emergency response of fire department vehicles

(d) Evaluation of the identified risks based upon the frequency and severity of these risks

(e) Development of an action plan for addressing each of the risks, in order of priority

(f) Selection of a means of controlling the risks

(g) Provisions for monitoring the effectiveness of the controls implemented

(h) A periodic review (annually) and required modifications to the plan

Responsibilities

This plan establishes a standard of safety for the daily operations of the department. The standard of safety establishes the parameters in which we conduct activities during emergency and nonemergency operations. The intent is for all members to operate within this standard of safety and not deviate from this process. We utilize a variety of control measures to ensure the safety and health of our members. These control measures include but are not limited to training, protective clothing and equipment, incident management, personnel accountability, and standard operating procedures. (See Table 5-1.)

The fire chief has responsibility for the implementation and operation of the department's risk management plan. The department's health and safety officer has the responsibility of developing, managing, and annually revising the risk management plan. The health and safety officer has the assignment of making modifications to the risk management plan based upon the prompt demand and severity of need based upon the monitoring of the process.

All members of the department have responsibility for ensuring their health and safety based upon the requirements of the risk management plan and the department's occupational safety and health program.

TABLE 5-1 Virginia Beach Fire Department Control Measures

Identification	Frequency/ Severity	Priority	Summary of Control Measures <u>O</u>ngoing or <u>A</u>ction required
Incident scene safety	Medium/high	High	1. *A*—Revise and implement department incident management system 2. *A*—Revise current policy on mandatory use of full personal protective equipment including SCBA 3. *A*—Evaluate effectiveness of the department's personnel accountability system and make needed adjustments 4. *A*—Enforcement of department procedures on mandatory use of PPE 5. *A*—Implementation of high-rise incident procedures. 6. *A*—Better utilization of the post-incident analysis process, which includes a written report for all personnel of significant incidents
Compliance with VOSH	Low/high	High	1. *A*—Compliance with 29 CFR 1910.134 "Respiratory Protection Program" (2 In/2 Out) 2. *A*—Develop a department compliance program for all mandatory VOSH programs
Health and wellness	Low/high	High	1. *A*—Implement the health and wellness strategic plan to improve the health maintenance, fitness, and wellness 2. *A*—Upgrade health maintenance program to include physical stress tests for all personnel 3. *O*— Continue to monitor participation and success with a physical fitness program
Health exposures	Medium/high	High	1. *A*—Provide annual retraining on infection control procedures 2. *A*—Update the City's Exposure Control Plan and the Infection Control SOP 3. *A*—Implement city and department policy and procedures for an occupational exposure to tuberculosis 4. *A*—Monitor the asbestos exposure control plan for exposures during emergency incident operations 5. *O*—Continue mandatory training and education program for HAZWOPER
Facility safety	Medium/high	High	1. *A*—Conduct a fire, safety, and health inspection of all fire department facilities 2. *A*—Conduct an inventory of all fire department facilities in conjunction with risk management 3. *A*—Monitor the infection control compliance within all fire department facilities
Vehicle accidents	Medium/high	Medium	1. *A*—Enforcement of state motor vehicle laws and department procedures relating to emergency response 2. *O*—Monitor nonemergency vehicle operations 3. *O*—Monitor individual members' driving records

TABLE 5-1 (*Continued*)

Identification	Frequency/ Severity	Priority	Summary of Control Measures *O*ngoing or *A*ction required
Tools and equipment	Low/medium	Medium	1. *A*—A review of the departments' 1996 Accident/Loss statistics indicate more accountability of lost/stolen/ damaged equipment 2. *A*—Implement loss reduction procedures 3. *O*—Maintain department equipment inventory
Strains and sprains	High/medium	Medium	1. *A*—Based on 1996 incidents, continue to monitor frequency and severity of incidents 2. *O*—Evaluate function areas to determine location and frequency of occurrence
Cuts and bruises	Medium/ medium	Medium	1. *A*—Require all supervisors to review use of protective clothing and equipment during emergency and nonemergency operations 2. *O*—Determine if protective clothing and equipment will reduce the number of incidents based on analysis
Safety and security	Medium/ medium	Medium	1. *A*—Awareness training on fire training, resource management, and fire administration for all members to be conducted on a company-level basis 2. *A*—Inspection and evaluation of crime prevention by city police at all facilities 3. *A*—Develop written procedures as necessary
Environmental hazards	Medium/high	Medium	1. *O*—Department compliance with 29 CFR 1910.120 2. *O*—Written procedures for the hazardous materials team and environment inspectors regarding illegal dumping of hazardous waste and hazardous waste sites
Protective clothing and equipment	High/high	Medium	1. *A*—Enforcement of department's procedures for use of personal protective equipment 2. *O*—Revise department policy and procedures on mandatory usage 3. *A*—Retraining and education of personnel on chronic effects of inhalation of by-products of combustion 4. *A*—Provide monitoring process of carbon monoxide (CO) levels at fire scenes, especially during overhaul
Environmental stress	High/low	Medium	1. *A*—Revise current department policy relating to Emergency Incident Rehabilitation 2. *O*—Evaluate and implement procedures for "weather extremes"
Financial	Low/high	Low	1. *O*—Maintain liaison with risk management, office of budget management, and city attorney

Monitoring Risks

The risk management program will be monitored on an annual basis. Recommendations and revisions will be made based on the following criteria:

(a) Annual accident and injury data for the preceding year

(b) Significant incidents that have occurred during the past year

(c) Information and suggestions from the Division of Risk Management

(d) Information and suggestions from department staff and personnel

(e) Evaluation of the risk management program by an independent source every three (3) years. Recommendations will be sent to the fire chief, the health and safety officer, and the occupational safety and health committee.

6

Role of the Company Officer and Safety Officer

Stephen N. Foley and Mary McCormack

This chapter focuses on the roles and responsibilities for fire fighter safety of individuals at supervisory levels, specifically of those who hold company officer or crew leader positions. At the tactical level of the incident, these supervisors manage specific tasks while also supervising the company/crew safety. They are responsible for ensuring that the operation is conducted safely, that the environment they are in is "as safe as possible," that a system of personnel accountability has been instituted, and that all personnel are equipped with and are properly using the appropriate personal protective ensemble (PPE).

The chapter also focuses specifically on the safety officer. The term *safety officer* is a generic title used to denote the department health and safety officer, the person responsible for the department occupational safety and health program. It is also used to denote the incident safety officer, the individual who, upon arrival at an incident, is assigned the duty of handling on-scene safety issues.

ROLE OF COMPANY OFFICER

Administratively, the level of supervision provided by the company officer or crew leader is probably one of the most difficult. Company officers may be supervising individuals who joined the fire department at the same time as they themselves did, or individuals who were promoted over their friends, or individuals who were transferred from a different company or station.

Supervision skills and leadership skills are distinct and different. Once a person is assigned the position of company officer or crew leader, he or she supervises personnel and performs tactical tasks at an incident. Company officers or crew leaders lead by example, operating safely, wearing the appropriate PPE, and "taking care" of their personnel. Crews or companies may be supervised by company officers who in the

Stephen N. Foley serves as the NFPA staff liaison for the Technical Committee on Fire Service Occupational Safety and Health.

Mary McCormack is executive director of the Fire Department Safety Officers Association (FDSOA).

past may not always have followed the rules. For such company officers to renounce that behavior and persuade their crew to do as they do now, not as they did in the past, is a difficult but necessary supervisory skill to learn. The fire department's safety standard operating procedures (SOPs) provide the supervisor with a tool for training personnel and for ensuring that the training is used at the incident scene.

Supervising personnel is challenging and is a skill for which little training is provided. Being a good fire fighter does not automatically make one a good supervisor. Ninety percent of a supervisor's job is dealing with people. Nowhere is this more important or evident than in the responsibility supervisors have for protecting the safety and health of their crew.

NFPA 1021, *Standard for Fire Officer Professional Qualifications*

NFPA 1021, *Standard for Fire Officer Professional Qualifications,* outlines four levels of requirements on different aspects of fire service occupational safety and health. Those striving to achieve these levels of professionalism must demonstrate their ability to meet these requirements. This standard, like all other NFPA standards, sets out the minimum requirements. In today's fire service, these include safety and health requirements that necessitate a broader approach to meet the intent of the standard. Such a broader approach includes understanding the concept of risk management, knowing how to interface with hospital and EMS personnel on an infectious or communicable disease issue, or working with outside agencies on accident or injury investigations.

Department's SOPs Dealing with Safety and Health Issues

Ensuring the safety and health of the crew is of paramount concern to a supervisor. Providing good leadership and setting a positive example demonstrate the commitment of the supervisor to this goal. Leadership skills, combined with experience and training, provide the supervisor with a base from which to implement standard operating procedures (SOPs). Included in the department's SOPs are those dealing with occupational safety and health issues, such as the wearing of protective clothing and equipment, personnel accountability, incident command, health and wellness issues, and critical incident stress (CIS).

The issue of CIS, for example, is one that includes resources from the fire department as well as from the community (e.g., the Red Cross, mental health professionals, and fire department membership organizations). The supervisor should recognize the signs and symptoms of CIS, know how to access resources to assist their personnel, and provide an ongoing dialogue with his or her company or crew.

Because of the environment in which the company officer or crew leader operates, a significant amount of training and education is required just to stay on top of these safety and health issues. Supervisors are not expected to know all the codes and standards referenced in this guide or the specifics of these codes and standards. As supervisors, however, they are responsible for company or crew safety and must lead by example. Following and enforcing the safety SOPs allows them to focus on the tactical assignments and to function as supervisors within the incident management system (IMS). Following the SOPs provides a degree of safety to all who are operating within the IMS. Deviating from the SOPs places individuals and the operation in jeopardy.

Additional Administrative or Operational Duties

In smaller and medium-sized fire departments, the supervisor may be assigned additional administrative or operational duties. In the area of fire service safety and health, these duties may include serving as a fire department safety and health officer, as a member of the fire department occupational safety and health committee, or as the incident scene safety officer. These assignments obviously require the company officer to take on additional roles and to carry more responsibility.

The health and safety officer's (HSO) responsibilities include a great deal of administrative work, data collection, and service as a liaison between personnel and various organizations. The qualifications and authority of the health safety officer are spelled out in NFPA 1521, *Standard for Fire Department Safety Officer* (see Figure 6-1).

The incident scene safety officer (ISO) assists the incident commander (IC) within the incident management system (IMS) as part of the command staff. In addition,

---

2-2   Qualifications of the Health and Safety Officer (HSO)

2-2.1 The health and safety officer shall be a fire department officer and shall meet the requirements for Fire Officer Level 1 specified in NFPA 1021, *Standard for Fire Officer Professional Qualifications*.

2-2.2 The health and safety officer shall have and maintain a knowledge of current applicable laws, codes, and standards regulating occupational safety and health to the fire service.

2-2.3 The health and safety officer shall have and maintain a knowledge of occupational safety and health hazards involved in emergency operations.

2-2.4 The health and safety officer shall have and maintain a knowledge of the current principles and techniques of safety management.

2-2.5 The health and safety officer shall have and maintain a knowledge of current health maintenance and physical fitness issues that affect the fire service members.

2-2.6 The health and safety officer shall have and maintain a knowledge of infection control practice and procedures as required in NFPA 1581, *Standard on Fire Department Infection Control Program*.

2-3   Authority of the Health and Safety Officer (HSO)

2-3.1 The health and safety officer shall have the responsibility to identify and cause correction of safety and health hazards.

2-3.2 The health and safety officer shall have the authority to cause immediate correction of situations that create an imminent hazard to members.

2-3.3 Where nonimminent hazards are identified, a health and safety officer shall develop actions to correct the situation within the administrative process of the fire department. The fire department health and safety officer shall have the authority to bring notice of such hazards to whoever has the ability to cause correction.

FIGURE 6-1   Qualifications and Authority of the Health Safety Officer
Source: NFPA 1521, *Standard for Fire Department Safety Officer*, 1997 edition

assistant ISOs who have technical expertise in a specific area—for example, technical rescue—may be used, as well. The ISO

- Serves as a member of the IC's command staff during incident scene operations
- Develops the safety plan as part of an incident action plan (IAP)
- Has the authority to terminate unsafe actions or operations at the incident scene
- Coordinates the actions of assistant safety officers as assigned at the incident scene
- Ensures that incident scene rehabilitation activities are addressed by the IC
- Assists on investigations as required by the health safety officer
- Ensures that a safety officer with the specificity of special operations is assigned for those specific incidents, for example, hazardous materials or technical rescue

The roles and responsibilities for the positions of health and safety officer and incident safety officer are outlined in NFPA 1500, *Standard on Occupational Safety and Health Program,* and more completely in the companion standards, NFPA 1521, *Standard for Fire Department Safety Officer,* and NFPA 1561, *Standard on Emergency Services Incident Management System.*

Incident Scene Safety Officer

As just discussed, officers assigned to serve as ISOs within the IMS discover that many responsibilities accompany this position. Serving as a member of the IC's command staff requires a defined set of skills. In recent years there has been a focus on providing training for persons assigned to the position of ISO (see Figure 6-2). State and provincial training academies, the National Wildfire Coordinating Group, as well as the U.S. National Fire Academy have developed courses for the ISO. In addition, some of these organizations have also sought accreditation to certify persons in this position. The Fire Department Safety Officers Association (FDSOA) is one of the leading associations pushing for training and certification in this area.

FIGURE 6-2   Using Different Positions Within the Incident Command System During Training Exercises Proving to Be a Valuable Tool in Training New Personnel
Source: Photo courtesy of Virginia Beach Fire Department

ROLE OF THE SAFETY OFFICER

Historical Background

In today's industrial environment, the term *safety* is an integral component of an organization's manufacturing or system process. The industrial safety person functions in a manner similar to that of the ISO in fire department operations. When the ISO arrives on-scene, he or she receives a briefing from the IC and then functions as the IC's eyes and ears for incident safety. ISOs handle on-scene safety issues, such as ensuring personnel accountability, ensuring the structural integrity of the building, and ensuring that personnel are assigned for rehabilitation. The ISO's responsibilities may vary from fire department to fire department.

When the position of ISO was first introduced in the fire service, the person filling that position was typically assigned to watch for signs of an imminent building collapse. Since then, the role of the ISO has been greatly expanded.

In 1970 the Occupational Safety and Health Act was signed into law. The law required the Occupational Safety and Health Administration (OSHA), a new organization set up as part of the U.S. Department of Labor, to develop and promulgate mandatory occupational safety and health standards.

NFPA Safety and Health Standards

In 1977 the National Fire Protection Association (NFPA) issued the first standard for the fire department safety officer. However, little notice was taken of it until 1987 when NFPA issued a new standard, NFPA 1500, *Standard on Fire Department Occupational Safety and Health Program,* and reissued the first standard as NFPA 1521, *Standard for Fire Department Safety Officer.*

Safety Officer's Responsibilities

Although the chief of department has the ultimate responsibility, the safety officer, whether in a career, a volunteer, or a combination department, has become essential in developing and managing the department's safety and health program.

In addition to being responsible for developing, managing, and monitoring the department's occupational safety and health program, the safety officer is also responsible for monitoring the risk management program. This program ensures that risks associated with department operations are identified and managed and the program is communicated through education and training of the members. The safety officer's other responsibilities include the following:

- Infection control
- Critical incident stress management
- Incident management safety
- Data collection and post-incident analysis
- Record management

The safety officer may also be called upon to assist with the other safety sections within the local governmental agencies.

Although the safety officer reports directly to the chief of department, he or she must also have a good working relationship with all the other officers and members of the department. Good communications skills, compassion, and mutual respect are skills and traits required of the safety officer. In larger fire departments, the safety officer also chairs the department's safety and health committee.

Fire Department Safety Officers Association (FDSOA)

Mission. In 1989, the Fire Department Safety Officers Association (FDSOA), an international association, was established for the purpose of addressing issues of safety for fire fighters. The association's mission is to promote safety standards and practices in the fire, rescue, and emergency services community. The FDSOA, designed to be a communications conduit for safety officers, publishes *Health & Safety for Fire and Emergency Services*. This monthly publication is a vehicle for safety officers and others involved with safety and health programs to share ideas by writing articles or by asking readers for answers to specific problems. The association also issues a monthly bulletin, entitled "Safety-Gram," to assist in training first responders.

The association has a membership base of 2,500 active safety officers. Through the educational programs it offers, people can exchange ideas with others who have the same problems and challenges. In addition, through the forum on the association's website, individuals can communicate directly with fellow safety officers throughout the world to improve the quality of fire fighter safety.

Certification Program. In 1999 the association became an accredited agency of the National Board on Fire Service Professional Qualifications. This accreditation enabled the association to begin a certification program for the fire department incident safety officer—fire suppression and to continue working on the health and safety officer certification program. Today there are over eight hundred certified ISOs across the United States. In early 2003, the association is expected to begin certifying the fire department health and safety officer.

SUMMARY

As individuals in supervisory positions, the company officer and crew leader are responsible for fire fighter safety. Their job is to ensure that the operation is conducted safely, that the environment in which they and their personnel are working is as safe as possible, that a system of personnel accountability has been instituted, and that all personnel under their supervision are properly equipped and are wearing appropriate PPE.

The safety officer is a generic title used to denote the person responsible for the department's occupational safety and health program as well as to denote the incident safety officer—that is, the individual who handles on-scene safety issues as part of the incident command staff within the incident management system. For additional information on the roles and responsibilities of those whose job it is to develop and manage the department's safety and health program and ensure fire fighter safety at the incident scene, see NFPA 1500, *Standard on Fire Department Occupational Safety and Health Program;* NFPA 1521, *Standard for Fire Department Safety Officer;* and NFPA 1561, *Standard on Emergency Services Incident Management System.*

Investigating Significant Injuries

Edward L. Stinnette

Investigation of a line-of-duty death or serious injury of a member of the fire department is one of the most difficult and important activities that we must conduct. This difficulty is compounded by the fact that the investigation must usually be conducted under extremely stressful circumstances and often under pressure for the rapid release of information.

The procedures for investigations outlined here can be applied to other situations, particularly accidents that result in serious injuries or in death. A "close call" should be interpreted as a warning to prevent the same situation from happening again. This will require supervisors to ensure appropriate control measures are in place.

The investigation of a line-of-duty death or serious injury is critical to improving the safety of all members of the department. The investigation must be a methodical effort to collect, analyze, and report, in an accurate and unbiased manner, the facts surrounding an incident.

The investigation of a line-of-duty death or serious injury may serve several different purposes. The Fairfax County Fire and Rescue Department will never be satisfied until we can be sure that we are doing everything in our power to prevent accidents, injuries, occupational illnesses, and line-of-duty deaths.

The goals of the investigation are to identify deficiencies in policy, procedures, and other actions that contributed to the incident and to make corrective recommendations to prevent this type of incident from occurring again.

OBJECTIVE OF THE INVESTIGATION

The objective is to determine the direct and indirect factors that resulted in a line-of-duty death or serious injury. The investigation must satisfy the requirements of the Public Safety Officer Benefits (PSOB) Program and other entitlements in the event of a line-of-duty death. These include but are not limited to the following:

Edward L. Stinnette is fire chief of the Fairfax County Fire and Rescue Department in Fairfax County, VA.

Source: This chapter is excerpted from the *Significant Injury Investigation Manual* prepared by the Occupational Health and Safety Division of the Fairfax County Fire and Rescue Department, Fairfax County, VA.

- Identifying inadequacies involving apparatus, equipment, protective clothing, SOPs, supervision, training, or performance
- Identifying situations that involve an unacceptable or unavoidable risk
- Identifying previously unknown or unanticipated hazards
- Identifying actions that must be taken to address problems or situations that are discovered
- Providing factual information to assist those involved who are trying to understand the events they experienced
- Ensuring that the incident and all related events are fully documented and evidence is preserved to provide for additional investigation at a later date
- Ensuring that the lessons learned from the investigation are effectively communicated to prevent future occurrences of a similar nature
- Providing the information to other groups that are involved in the cause of fire fighter occupational safety and health

INVESTIGATION TEAM AND PROCESS

Investigation Team

A thorough investigation will usually require at least five individuals and may involve a larger team. Additionally, technical specialists and the professional standards officer will be utilized as needed. The administrative division assistant chief will maintain a list of persons who can be immediately activated when an incident occurs. Designated team members should respond to the scene of the incident to begin the investigative procedures. The team members should be immediately reassigned from their regular duties to devote their full efforts to the investigation.

Figure 7-1 illustrates the organization of a hypothetical investigation team.

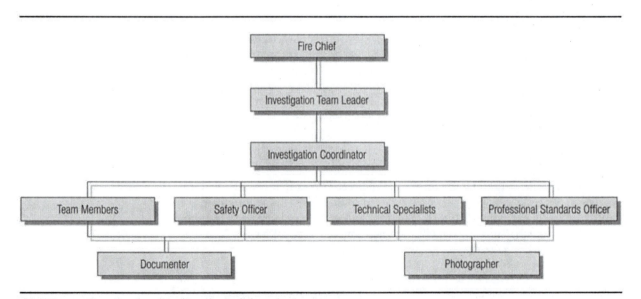

FIGURE 7-1   Organization of a Hypothetical Investigation Team

Team Leader. Of utmost importance in appointing a team leader is to delegate the necessary authority to conduct a complete and thorough investigation. A team leader who holds command or management level rank can usually function more efficiently in gaining cooperation and coordinating team efforts. The individual should be respected for his or her expertise, impartiality, and conscientious work. *No other individual should have the authority to interfere with the investigation.*

Investigation Coordinator. The investigation coordinator serves as an assistant to the investigation team leader. Appointment of an investigation coordinator is optional but highly recommended for larger investigations. The investigation coordinator should have management experience and be involved with fire and rescue department operations.

Safety Specialist. The safety specialist should be a safety officer from the occupational safety and health division. The safety specialist should have a good working knowledge of federal, state, and local laws and of OSHA and NFPA regulations and standards.

Team Members. The core of the investigation team members is from the fire department. The members should have both technical knowledge and experience in various aspects of department operations. These areas include training, incident operations and command, and accountability.

Technical Specialists. Individuals with expertise in a specific area are needed to evaluate aspects of the investigation. Specialists in the field of protective clothing, vehicle maintenance, law enforcement, communications, photography and videography, and medical evidence may be called upon as dictated. The use of technical experts greatly enhances the impartiality of the investigation. A police detective should be a member of every investigation team. The team leader should not hesitate to involve experts in the investigation when the technical demand exceeds the team's capabilities.

Investigation Process

A thorough investigation will require both time and effort. It is important to discover, identify, research, and fully document every factor or potential cause. The investigation should focus on factual information. It should present the facts of what happened, identify the factors, and recommend appropriate corrective actions. In many cases there will be conflicting theories and opinions about the incident. There may also be a number of very different accounts from witnesses and individuals who were involved. The investigation should follow up on every lead or theory to discover the actual facts, as precisely as they can be determined.

The investigation team may be placed in a difficult position of investigating the actions of friends, co-workers, and superior officers. There may be internal pressure to find a particular individual or certain action responsible for the death or injury. There may also be the temptation to omit or cover up details in order to protect the reputation of a specific individual or the department.

When a serious incident occurs, emotional reactions are natural, and the investigation process can magnify them. The investigation must always attempt to separate facts from emotions and opinions and be limited to finding the facts and developing recommendations.

The investigation will require a high level of cooperation between the investigative team and other agencies and organizations that may be involved in investigating or seeking information on the incident. This may include organizations that have a statutory authority or responsibility to investigate the incident and others that have legitimate reasons to be involved or to be interested in the results. There may also be organizations that are requested to assist the investigation team. The best policy is to be cooperative with other agencies that have a recognized reason to be involved in the investigation.

The investigation team shall be the authority having jurisdiction over the fire department's internal investigative process. If the investigation surrounds a fatality, the police homicide section will be the lead agency for that aspect of the process. A member of the investigative team should be on the scene before fire department operations are completed and should retain control of the scene as long as is necessary to conduct the investigation. If it is not a fire incident, control of the scene may fall within the jurisdiction of another agency and the investigation team will have to seek their cooperation to complete its own scene research.

A fire cause investigation may be carried out concurrently with the line-of-duty investigation. If there is evidence of arson or other criminal acts, the situation will become much more complicated. The investigation of the safety factors involved in the incident must continue, while a high level of coordinating is provided with fire investigation and law enforcement investigators. The fire department should retain custody of the scene until each investigation team has completed its work.

Conducting the Investigation

The initial stage of the investigation will generally require several hours and should be conducted according to a methodical plan.

Securing the Scene.   The scene of the incident should be secured and guarded; only those individuals who have a specific reason to enter should be allowed inside the perimeter. An officer and as many members as are necessary should be assigned to secure the scene. Police assistance may be necessary to establish and maintain scene security. The sooner that isolation is implemented, the easier it will be to investigate the scene and to account for any disruptions of the physical evidence.

A log should be maintained of all personnel entering and exiting the secured area. Evidence, apparatus, hose lines, or equipment should not be moved until the investigation team has recorded its position and conferred with police or arson investigators. The only reasons to violate this rule would be to provide medical treatment in an attempt to save a victim or to control a fire that could destroy the evidence. The scene should be maintained until all physical evidence has been documented, photographed, and measured.

Seizing Evidence.   All items that could have a bearing on the investigation should be impounded and protected until they can be turned over to the investigation team. This will include any items the individual was wearing, using, or operating.

Physical evidence should be handled in the same manner as evidence from an arson investigation or criminal investigation. The incident commander (IC) should immediately assign someone to take custody of any items that are removed from the secured area and to turn them over to the investigation team. Any movement of evidence should be noted and recorded, such as the removal of personal protective equipment, work uniform, and/or breathing apparatus to treat fire fighters.

During the investigation, the investigators are responsible for locating, collecting, identifying, storing, examining, and arranging for the testing of physical evidence that may prove or disprove a particular fact or issue. The investigator must be thoroughly familiar with chain-of-custody procedures and accepted methods of processing physical evidence. The law enforcement or fire investigation technical specialist can assist with this task.

Documenting Physical Evidence.   There are certain items at the incident scene that should always be documented. These items are presented in Figure 7-2. Although the investigators are looking for physical evidence, it is also advisable to look for what is missing. What is not found at the incident scene can be as significant as what is there.

The goal of describing the incident scene is to record the position of people, tools, apparatus, and elements of the physical environment. The scene and evidence should be diagramed, photographed, and videotaped in the same manner that a crime scene would be documented.

PHYSICAL EVIDENCE DOCUMENTATION CHECKLIST

_____   Location and position of dead or injured persons

_____   Position of hose lines

_____   Location and position of apparatus

_____   Location of command post

_____   Location of tools and equipment

_____   Location of pieces broken off from equipment or tools

_____   Location of windows, doors, stairs, and ventilation openings (including details of whether doors and windows were open/closed or broken/intact)

_____   Location of any furniture or obstacles

_____   Areas of debris from roof or floor collapse

_____   Incident management structure at time of incident

_____   Incident commander and sector officers with their locations at time of incident

_____   Environmental conditions (night, wind, rain, snow)

_____   Protective clothing and equipment (secured as recovered)

_____   Scratches, gouges, dents, or breakage related to fire fighter activities

_____   Any other items that appear significant

_____   Building documents including past inspections

FIGURE 7-2 Sample Physical Evidence Documentation Checklist

Examining Physical Evidence.   Once collected, physical evidence should be examined and tested by an independent laboratory or other testing facility to determine their compliance with their recognized standards.

Identifying Witnesses.   It is often impossible for the investigation team to interview all of the witnesses at the scene or immediately after the incident. The immediate priorities should be to obtain essential information from individuals who were directly involved and to identify witnesses for later follow-up.

Conducting Interviews.   Full interviews should be conducted with every fire department member involved in the event as well as any civilian witnesses. At a major incident, this may have to be confined to those who were at the scene at the time of the incident or who were in any way involved with the victim before or during the event. The information gathered during the interview should always be recorded on paper and with a tape recorder.

Anyone who has information that could be significant should be encouraged to inform the investigation team. Every contact should be interviewed, including members of the public.

Interviews should be open, honest, and nonjudgmental of the interviewee. The goal is simply to share information about the incident. It is essential that the investigator establish trust with all witnesses. It must be stressed that the purpose of the interview is to prevent future incidents and not to conduct a witch-hunt.

Anyone who makes a statement reported in the news media should be located and interviewed. News media statements often confuse the issues in the early stages of an investigation; finding the person who made a statement is usually the best way to determine its accuracy.

The team should obtain and review copies of all news broadcasts and published accounts of the incident. The reporters themselves should be interviewed, if their reports suggest some factor not consistent with the information found by the team.

The questions in the sample witness interview questionnaire shown in Figure 7-3 should be used as a guide and adjusted to match the events of the incident.

Developing a Time Line.   The compilation of records, radio tapes, and other data should allow the investigative team to establish a basic time line for the incident. The time line establishes the sequence of events chronologically, sometimes to the second. Additional information should be added to the time line as it is obtained, until the time line can be used to fully describe who did what, who saw what, at what location, and at what time. Developing a time line is one of the basic building blocks of an investigation process.

In the establishment of a time line, it is important to synchronize the time base for different records. Misleading information may result if times are compared from different sources, assuming that the clocks were synchronized at the time of the incident. The investigation team should verify the times that are recorded from a verifiable simultaneous event and apply the appropriate correction factor to all other time measurements.

Obtaining Records of the Incident.   All of the following information should be reviewed to determine if there is anything to suggest contributory causal factors or to fill in missing information:

- Records of all telephone calls (including any cell phone conversations)
- Recording of all radio traffic

WITNESS INTERVIEW QUESTIONNAIRE

Witness Identification

1. What is your legal name? _____

2. Do you have a nickname or other name by which co-workers call you?_____

3. What is your home address and telephone number? _____

4. What is your current work assignment including shift? _____

5. Do you have any physical problems that affect how you hear or see?_____

6. Who is your supervisor?_____

7. How long have you been at your current assignment?_____

8. How long have you been with the Department? _____

Time and Place of Incident

1. What time did the incident occur?_____

2. Describe the incident scene. _____

   _____

   _____

3. Describe the area surrounding the incident scene. _____

   _____

   _____

4. Were there any environmental factors that contributed to or complicated the

   incident? _____

5. Describe the building design and construction. _____

   _____

   _____

Incident Description

1. Describe in detail what you saw and heard at the incident scene from when

   you first arrived until the end of the incident. _____

   _____

   _____

(Page 1 of 5)

FIGURE 7-3   Sample Witness Interview Questionnaire

2. Draw a schematic of the incident scene.

3. When did you first realize that something was wrong? _____

4. What work tasks were you assigned to do? _____

_____

5. Did you complete the assigned work tasks? _____

6. What did you hear? _____

_____

7. What did you see? _____

_____

8. What commands were you given during the incident? _____

_____

9. Who gave them to you?_____

_____

_____

10. Were you properly trained and equipped to carry out those commands?_____

Interactions with Other Witnesses or Interviewers

1. Have you spoken with anybody else about the incident? _____

2. Tell me about the conversation. _____

_____

_____

3. Have you given any other statements about this incident?_____

4. To whom have you given other statements? _____

5. When did you make the statement? _____

6. Was anybody else present while you were making the statement?_____

(Page 2 of 5)

FIGURE 7-3   (Continued)

7. Were there any notes or records made of the statement? _____

8. Have you prepared any written statements about the incident? _____

9. To whom did you give the statement?_____

10. Do you have a copy of the statement? _____

11. Have you made any maps, sketches, or diagrams or taken any

    photos of the incident scene?_____

12. Do you know anyone who has? _____

When a Tool or Piece of Equipment Fails

1. Were you aware that the _____ was not functioning properly at the

   time of the incident? _____

2. Who else knew about this?_____

3. Did you or anyone else take actions when you noticed the defect? _____

4. Were any of the observations or actions documented in the station log or by repair requests?

   _____

5. Has the _____ ever failed before? _____

6. Please tell me about the other times that _____ has failed.

   _____

   _____

7. What corrective actions were taken to rectify this problem?_____

   _____

   _____

8. Were there any warning signs this time of the failure? _____

9. When was the last time preventative maintenance was performed? _____

10. How often was the _____ typically used? _____

Maintenance, Inspection, and Testing

1. What is the regular maintenance program for the _____?

2. Please describe the cleaning and maintenance procedures. _____

   _____

(Page 3 of 5)

FIGURE 7-3   (*Continued*)

3. What type of preventative maintenance is performed? _____

_____

4. What and where are the maintenance records? _____

5. Have any defects been noted? _____

6. Who is responsible for maintenance or daily inspections? _____

When Fire Is Involved

1. When did the fire start? _____

2. Where did the fire start? _____

3. When did you first notice the fire? _____

4. Who notified the fire and rescue department? _____

5. How long did it take for the first unit to arrive? _____

6. When was the fire declared under control? _____

7. When was the incident terminated? _____

8. What caused the fire? _____

9. Were there any unusual smells? _____

10. What color was the smoke? _____

11. What part of the building was affected by fire? _____

12. What area was affected by smoke? _____

13. What area did water or extinguishing agents affect? _____

14. Were there any sprinklers in the building? _____

15. Did they function properly? _____

16. Did all escape devices function properly (stairs, ladders,

    fire escapes, exit doors/signs)? _____

17. Describe the structural integrity of the building. _____

_____

_____

18. Were there civilians in the building? _____

(Page 4 of 5)

FIGURE 7-3   *(Continued)*

19. Why were you in the building? _____

20. Were there any hose lines advanced? _____

21. Were there any ground and/or aerial ladders placed?_____

22. Where were they placed? _____

23. What was your initial role in the response? _____

When Someone Has Fallen

1. Where did you/they fall?_____

2. What was the initial position before falling?_____

3. Where did you/they land?_____

4. What were you/they doing when the fall happened?_____

5. Were there any sounds or cries prior to the fall? _____

6. What injuries were sustained in the fall? _____

7. What type of footwear was used? _____

8. Was the footwear in good condition? _____

9. What was the condition of the floor? _____

Departmental Standard Operating Procedures (SOPs)

1. What departmental SOPs applied to the incident? _____

    _____

2. Please describe the SOPs. _____

    _____

3. Is this/these SOPs generally followed? _____

4. When have this/these SOPs not been followed? _____

    _____

5. Compare what actually happens in the field to what the SOPs say._____

    _____

    _____

(Page 5 of 5)

FIGURE 7-3   *(Continued)*

- Photographs and/or videos by the fire department, media, or public
- Written reports by the fire department
- Other pertinent information

Researching Documents. All existing departmental SOPs, training materials, and similar sources of guidance that would apply to the situation should be reviewed to determine

- How the situation "should" have been handled
- Whether or not it was handled in the expected manner
- Whether or not this would have had an impact on the outcome

Records should be examined to determine if the individuals involved had received the proper training in the relevant topics.

All applicable NFPA standards, ANSI standards, OSHA regulations, and similar information that could relate to the events should also be studied. NFPA annual reports on fire fighter death and injuries should be consulted to determine if similar situations have occurred in other departments. The conclusions from those reports should be compiled; and, if possible, the final reports from those incidents should be obtained.

Where equipment or apparatus is involved, specifications and maintenance records should be obtained. Operators should be asked if any problems were previously noted and a determination should be made if required inspections and repairs had been completed on schedule. Time should be taken to interview the maintenance crews.

Using Outside Assistance. There are several situations that will require the assistance of qualified experts. Assistance is available in many different areas. If the needed expertise is not available within the fire department, finding the best individuals to assist the team in specific areas or to be part of the entire investigation is an excellent investment. If an incident has become extremely controversial, it may be advisable to have a recognized independent investigator participate in the investigation or review the evidence to develop an independent report.

Obtaining Legal Advice. Legal issues will involve nearly every aspect of investigation of a line-of-duty death or significant injury. Where potential criminal action is a possibility, the safety investigation should be independent but must be coordinated with the appropriate law enforcement agencies. Issues of potential liability, including product liability and possible violation of OSHA laws, will be a consideration in almost every case. These factors should not be allowed to restrict the investigation, but it is advisable to have the report reviewed by legal counsel before it is released.

Written Report

The process of investigating a death or significant injury generates many facts and much speculation. The investigation report provides a condensation of relevant facts about the incident. When writing the report, investigators must keep in mind that the objective of the investigation process is to discover exactly what happened and determine how to prevent similar occurrences. All items in the report must be precise, accurate, and complete. The written report is divided into the following sections: incident summary, facts, analysis, conclusions, and recommendations.

The report should be presented by the team leader to the fire chief as a completed document at a meeting with all team members present. The team leader should present an overview of the report, including all conclusions and recommendations, using audiovisual aids to illustrate the presentation. The team should be prepared to address any questions and/or concerns from the chief and his or her staff. The decision to release the final report will be determined through discussions between the fire chief and city or county attorney's office.

The report will be presented to the fire department's Occupational Safety and Health (OSH) Committee. The committee should review the entire report, paying particular attention to the recommendations to prevent further occurrences of a similar nature. As a representative body, the OSH Committee adds credibility to the investigative process and the final report.

The report should be presented in a special session to the members involved with the incident. In most cases this will bring the incident to closure for those individuals. The report should then be released to the department. Every department member should use the final report or a presentation of its major points.

The release of the completed report makes it a public document, accessible to the news media and any other interested party. If there is a known media interest in the report, copies should be made available to those who have requested it.

When appropriate, copies of the report should also be sent to organizations that are involved in fire fighter safety and health (i.e., NFPA, IAFF, IAFC, etc.), as well as to neighboring jurisdictions if requested.

SUMMARY

The investigation of a line-of-duty death or serious injury is critical to improving the safety of all members of the fire department involved. The objective is to determine the direct and indirect factors that resulted in the death or serious injury, with the purpose of preventing similar accidents, injuries, occupational illnesses, and line-of-duty deaths.

The investigation team consists of a team leader, an investigation coordinator, the safety officer, technical specialists with expertise in specific areas, and team members from the fire department. The investigation process focuses on factual information. The investigation team should try to separate facts from emotions and opinions and should present the facts of what happened and recommend appropriate corrective action.

The steps involved in conducting the investigation include securing the scene, seizing evidence, documenting and examining physical evidence, identifying witnesses and conducting interviews with them, developing a time line of the events, obtaining records and researching documents, and obtaining outside expert assistance or legal advice, if necessary. The culmination of the investigation is a written report, which is a condensation of the relevant facts about the incident, as well as an analysis of these facts, the conclusions reached, and a recommendation for future action. Additional information and resources are provided in Chapter 15, "Line-of-Duty Death Resources for Fire Departments."

Case Histories of Fire Fighter Fatalities and Injuries

Included in NFPA 1500, *Standard on Fire Department Occupational Safety and Health Program*, 2002 edition, is a section in Chapter 8 on post-incident analysis. For some hardened and scarred fire service veterans, this analysis used to be called a post-incident critique, with sessions sometimes turning into a finger-pointing exercise. Today the emphasis is on who did what right, on the success of the incident, and on ensuring that no one was injured.

Part III opens with Chapter 8, "Lessons Learned from Investigating Incidents," which discusses the use of case histories in the fire service—that is, what lessons can be learned from investigating and analyzing incidents.

Chapters 9 through 13 provide case history summaries of recent incidents involving fire fighter fatalities. Chapter 9, "Protecting Emergency Responders: Lessons Learned from Terrorist Attacks," provides a summary report on the September 11, 2001, tragedy at the World Trade Center, as excerpted from conference proceedings published by the Science and Technology Institute, which is sponsored by the National Science Foundation and managed by Rand Corporation.

Chapter 10, "Residential Fire, Keokuk, Iowa, December 22, 1999," and Chapter 11, "Fire Fatalities, Marks, Mississippi, August 29, 1998," are both NFPA investigation report summaries. Chapter 12, "Six Career Fire Fighters Killed in Cold Storage and Warehouse Building Fire, Worcester, Massachusetts," is the summary of a NIOSH Fatality Assessment and Control Evaluation Investigative Report. Chapter 13, "South Canyon Fire Investigation (Storm King Mountain)," is the executive summary of the report of the South Canyon Fire Accident Investigation Team under the auspices of the National Interagency Fire Center in Boise, Idaho.

Information and material gleaned from these investigations have assisted NFPA and other standards-making organizations in the development and revisions to fire service health and safety standards. We may wonder, however, to what extent the fire service has learned from these lessons.

8

Lessons Learned from Investigating Incidents

Robert F. Duval

In reviewing case histories and official reports of incidents where fire fighters are killed or injured, we can analyze the incidents to extract the lessons learned. And there is always a lesson to be learned. Whether the incident is of a massive scale (the terrorist attack on the World Trade Center in New York), of a large scale (e.g., Worcester, Massachusetts, cold-storage building fire or the Storm King Mountain wildland fire in Colorado), or of a local scale (e.g., Marks, Mississippi, florist shop fire or Keokuk, Iowa, residential fire), each incident contains aspects that can be analyzed and used to educate responders in the hope that such a tragedy is not repeated.

MASSIVE SCALE INCIDENT LESSONS

Each incident is unique and may present lessons that will pertain only to incidents of similar magnitude and scope. Others may offer common lessons that will pertain to everyday operations. Take, for example, the World Trade Center attacks presented in Chapter 9. The RAND/NIOSH report, entitled "Protecting Emergency Responders: Lessons Learned from Terrorist Attacks," reviews in detail the problems encountered by first responders at the World Trade Center site in 2001 and compares them to the problems encountered at the Oklahoma City Bombing incident of 1995, as well as the anthrax threat responses in the fall of 2001. These problems included unknown hazards (biological, chemical, environmental, as well as physical) encountered by the responders.

Standard Personal Protection Equipment (PPE) Limitations

The limitations offered by standard personal protective equipment (PPE) weighed heavily on the response of rescuers and support personnel. Fire fighter's PPE works well under structural fire conditions, but it is not well suited for long-term search and rescue operations over rubble and in the face of hazards presented by that environment. Police and EMS responders are not adequately equipped or protected for operating in that type of environment.

Robert F. Duval is senior fire investigator and New England regional manager at the National Fire Protection Association.

Respiratory and Vision Protection Limitations

Respiratory protection was of paramount importance at the scene of the World Trade Center attacks. However, the availability of respirators was limited in the early stages of the incident. Vision protection was another area where equipment was limited in the early stages. These limitations can be attributed to the magnitude of the incident and the number of responders.

Operating with an unknown hazard, such as with the anthrax threats, poses the dilemma of what type of protective equipment is needed. Many first responders had to reevaluate their response to such threats in the days following September 11, 2001. The events of late 2001 brought to the forefront the threats and hazards posed by terrorism and/or hazardous materials responses. These are certainly unique lessons and do not compare to more common-type incidents such as structure fire or wildfire incidents.

INCIDENT MANAGEMENT

One of the common threads in comparing the World Trade Center and Pentagon incidents to other incidents is incident management. Whether the incident is as massive as the World Trade Center and Pentagon incidents, or of a much smaller scale, such as the florist shop incident in Marks, Mississippi, the reasons for using the incident management system are basically the same, that is, accountability, safety of the responders and the public, and scene management, as set forth in NFPA 1561, *Standard on Emergency Services Incident Management System*.

Personnel Accountability

In looking at incidents other than the World Trade Center attacks, we can see that the use of the IMS plays a major role in the lessons learned from each incident. In the Keokuk, Iowa, and Marks, Mississippi, incidents, a lack of an established IMS led to a series of events that led to fire fighter fatalities. This lack of command structure early in the incident led to a lack of accountability of the personnel in the structure.

Keokuk, Iowa, Incident.   In the Keokuk, Iowa, incident (Chapter 10), the initial incident commander (the assistant chief) was immediately confronted with a rescue situation and was unable to assume command. When the chief arrived on the scene, he too was confronted with a rescue situation by being handed a child who had just been removed from the burning house and was in respiratory arrest. The chief responded by leaving to bring the child to the nearby hospital prior to the arrival of EMS units. In his absence, no IMS was established and in turn no accountability system was developed to track the personnel within the structure. Once the chief arrived back at the scene, he had no way to account for the three fire fighters in the now well-involved building. Repeated attempts to reach them by radio and physical contact were unsuccessful. The three fire fighters perished in a rapid build-up of heat and fire gases within the building. They had been attempting to rescue three small children, who also died in the fire.

Marks, Mississippi, Incident.   In the Marks, Mississippi, incident, what started out as a relatively small fire against the rear of a building ended with two fire fighters killed in two separate incidents on the fireground. Again a lack of an established IMS led to a series of events resulting in the deaths of the two fire fighters. A command structure was not established early into the Marks incident. What appeared to be a small fire in the rear of a main street block of businesses quickly took a turn for the worse when the fire extended into a concealed attic space. The rapid change in tactics from extinguishment of a small fire to extinguishment of a hidden fire in a much larger block of buildings resulted in a lack of accountability of fire fighters now battling a fire on several fronts.

As the fire was discovered to have spread into the adjoining block of stores, a fire fighter was sent to examine the conditions on the roof and to prepare to ventilate the attic. His partner, who was not properly equipped for the task, went to obtain breathing apparatus.

Before the partner could return to the fire fighter on the roof, he was given another assignment and was sent to the front of the building. The fire fighter sent to the roof fell through a weakened roof structure into the rear of the burning store. His fall was not noticed by anyone on the fireground. The fact that he was missing was not discovered for an extended period of time because all attention was focused, at the time, on a missing fire fighter in the front of the store.

As the fire attack was directed at the front of the store, two fire fighters entered with a hoseline to locate and confine the fire that had now spread into the store. As they entered the front door, an unknown event occurred, causing one of the fire fighters to lose contact with the hoseline and forcing the other fire fighter out the door onto the sidewalk. A search began for the fire fighter lost inside the building. Additional resources were requested to respond. Eventually, the fire fighter was recovered not more than 24 feet (7.3 m) from the front door. It was at this point that it was determined that anther fire fighter (the fire fighter who had been on the roof) was missing. His body was eventually located near where he had fallen through the roof.

In both of the Keokuk and Marks cases, the lack of accountability of all responders resulted in fire fighters working unsupervised in hazardous environments.

Worcester, Massachusetts, Incident. In the case of the Worcester, Massachusetts, cold-storage and warehouse building fire (Chapter 12), accountability became an issue as the interior layout of the building and the built-in hazards the building construction posed came into play. In an almost windowless building, fire fighters were faced with a hidden fire within a mazelike interior. Within ten minutes of the fire department's arrival, a report of two homeless people possibly being in the building led to an extensive search of the building. During the search, two fire fighters from the rescue company became disoriented on an upper floor in the building. As additional companies were deployed to search for the missing fire fighters, four other fire fighters became trapped in the confusing interior of the building under deteriorating conditions. All six of the trapped fire fighters perished as fire conditions drove the remaining units from the building.

Accountability of the many units on the scene in Worcester became difficult due to factors such as the size and layout of the building and the multiple units operating on several levels within the structure.

Size-up and Risk Management

Another factor that often contributes to fire fighter injuries and fatalities is a lack of or difficulty in sizing up an emergency scene. Size-up is a continuous process that begins with the dispatch and continues until the incident is under control. The size-up of an incident includes many factors that have to be evaluated. The initial incident commander and every subsequent incident commander must perform a risk-versus-benefit analysis of the emergency scene. In each of the outlined incidents, size-up and risk-versus-benefit analysis played a major role. Size-up determines the strategy and tactics to be used in stabilizing the incident.

Structural-Type Incidents.   At the World Trade Center incident, the initial incident commanders were faced with a major incident involving fire and structural damage in a high-rise building. Their early priorities included evacuating thousands of occupants as well as getting a better idea of the extent of the damage caused by first one and then two planes crashing into the towers. In the Worcester incident, the incident commanders were confronted with a hidden fire in a windowless building and then a report of people trapped inside. In the Keokuk incident, the first arriving officer was faced with an immediate rescue need that precluded him from conducting a further size-up and risk-versus-benefit analysis. In the Marks incident, the incident commander was faced with a rapidly changing fire condition that resulted in two fighters becoming trapped in two different locations at the same incident. The focus shifted from the extinguishment of a small fire to a major incident. Initial and ongoing size-up of the incident failed to discover the fire spread into the adjoining attic space.

Wildfire Incident.   In the case of the Storm King Mountain wildfire incident (Chapter 13), where 14 fire fighters died, size-up covers different components from those covered in a structural-type incident. Commanders in a wildfire situation have factors unique to wildland fire fighting that must be considered during the size-up phase. In the Storm King Mountain fire, changing weather conditions, terrain, and fire development played a role in the fire roaring up the steep slope and trapping the 14 fire fighters.

Being able to recognize signs of impending problems such as collapse, backdraft, flashover, or wildfire "blow-up," comes with training and experience. Recognizing that a dangerous situation exists or is about to get worse can allow fire fighters or incident commanders time to react and remove themselves or other responders from a dangerous situation. Risk management during emergency operations is addressed in NFPA 1500, *Standard on Fire Department Occupational Safety and Health Program.*

Resource Management

Obtaining and maintaining adequate resources at the scene of an emergency is necessary for a safe and efficient operation. These resources include personnel, apparatus, and equipment. When a fire attack or a rescue attempt is conducted, sufficient standby personnel and equipment are needed to provide support for the operation. To properly manage risk at an emergency scene, incident commanders must have sufficient resources at their disposal. NFPA 1500 addresses the allocation of resources in the initial moments of an operation. In both the Keokuk and Marks incidents, the allocation of resources played a major role.

Keokuk Incident.   In the Keokuk incident, the initial group of four fire fighters was quickly overwhelmed by the immediate rescue situation confronting them upon their arrival at the early morning residential fire. Three of the four fire fighters were deployed in the rescue effort, while the fourth secured a water supply one block remote from the fire scene. When two additional personnel arrived (the chief and another fire fighter), they were both deployed to assist with the rescue and rapidly spreading fire condition, leaving no one to provide a backup.

Marks Incident.   In the Marks incident, the initial resources were able to extinguish the fire in the rear of the store, However, when the fire was found to have spread to the adjoining building, the supply of air cylinders on scene had been nearly depleted. Then when a rescue effort had to be mounted to locate the fire fighter at the front of the building, additional units had to be called to the scene to assist.

SUMMARY

When a fire fighter fatality or injury occurs, there are always lessons to be derived from the incident. Unfortunately, many of these lessons are repeated over the course of several incidents and years. These repeated lessons include shortcomings in incident management, personnel accountability, size-up and risk management, and resource management. As George Santayana so aptly stated, "Those who cannot remember the past are condemned to repeat it." The object is to learn from these tragic incidents to avoid repeating them.

Protecting Emergency Responders: Lessons Learned from Terrorist Attacks [Excerpts]*

On December 9–11, 2001, a conference was held in New York City that brought together individuals with experience in responding to acts of terrorism. The purpose of the conference was to hear and document the firsthand experiences of emergency responders regarding the performance, availability, and appropriateness of their personal protective equipment as they responded to these incidents. The meeting considered the responses to the September 11, 2001 attacks at the World Trade Center and the Pentagon; the 1995 attack at the Alfred P. Murrah Federal Building in Oklahoma City, Oklahoma; and the emergency responses to the anthrax incidents that occurred in several locations through autumn 2001. The conference was sponsored by the National Institute for Occupational Safety and Health of the U.S. Centers for Disease Control and Prevention, which also arranged for RAND to organize and conduct the conference and prepare this report.

This report presents a synthesis of the discussions held at the December meeting. It is intended to help federal managers and decision makers

- Understand the unique working and safety environment associated with terrorist incidents
- Develop a comprehensive personal protective technology research agenda
- Improve federal education and training programs and activities directed at the health and safety of emergency responders

---

*This chapter consists of excerpts from conference proceedings, *Protecting Emergency Responders: Lessons Learned from Terrorist Attacks*, by Brian A. Jackson, D. J. Peterson, Kames T. Bartis, Tom LaTourrette, Irene Brahmakulam, Ari Houser, and Jerry Sollinger. The conference was organized and the proceedings published by the Science and Technology Institute. The Institute is a federally funded research and development center sponsored by the National Science Foundation and managed by RAND. For additional information or to download a copy of the full 89-page report, see http://www.rand.org/scitech/stpi. For illustrative purposes, photographs in this chapter have been added from another source. They are not part of the original report.

The report should also help state and municipal officials, trade union leaders, industry executives, and researchers obtain a better understanding of equipment and training needs for protecting emergency workers.

EXECUTIVE SUMMARY

Just as it has for the nation as a whole, the world in which emergency responders work has changed in fundamental ways since September 11, 2001. Members of professions already defined by their high levels of risk now face new, often unknown threats on the job. At a basic level, the September 11 terrorist events have forced emergency responders to see the incidents they are asked to respond to in a new light. At the World Trade Center, 450 emergency responders perished while responding to the terrorist attacks—about one-sixth of the total number of victims. Hundreds more were seriously injured. In this light, the terrorist events are also forcing emergency responders to reconsider the equipment and practices they use to protect themselves in the line of duty.

Preparation is key to protecting the health and safety of emergency responders, and valuable lessons can be learned from previous responses. To this end, the National Institute for Occupational Safety and Health (NIOSH) sponsored and asked the RAND Science and Technology Policy Institute to organize a conference of individuals with firsthand knowledge of emergency response to terrorist attacks. The purpose of the conference was to review the adequacy of personal protective equipment (PPE) and practices, such as training, and to make recommendations on how the equipment and practices worked and how they might be improved. Attendees included persons who responded to the 1995 attack on the Alfred P. Murrah Federal Building in Oklahoma City, the September 11 attacks on the World Trade Center and the Pentagon, and the anthrax incidents that occurred during autumn 2001. They represented a wide range of occupations and skills: firefighters, police, emergency medical technicians, construction workers, union officials, and government representatives from local, state, and federal agencies. The conference was held December 9–11, 2001, in New York City, and this report synthesizes the discussions that took place there.

NEW RISKS, NEW ROLES FOR EMERGENCY RESPONDERS

Although the terrorist incidents shared some characteristics with large natural disasters, the NIOSH/RAND conference participants highlighted ways in which those incidents posed unique challenges. They were large in scale, long in duration, and complex in terms of the range of hazards presented. As a result of these characteristics, these events thrust responders into new roles for which they may not have been properly prepared or equipped. The themes of scale, duration, and range of hazards were repeated frequently during the discussions at the conference because they were seen as having critical implications for protecting the health and safety of emergency responders—during both the immediate, urgent phase and the sustained campaign phase of the responses.

The September 11 terrorist incidents were outside for their large scale—in terms of both the damage incurred and the human and material resources needed to respond. Conference participants spoke extensively about the difficulty of conducting search and rescue, fire suppression, and shoring and stabilization operations, as well as hazard monitoring. Responses were hampered by collateral developments, in particular the grounding of commercial air transport, which slowed the implementation of command and logistical support infrastructures.

The responses to the terrorist attacks involved days and weeks of constant work. At the World Trade Center, an initial urgent phase persisted for several days and then gradually transitioned into a sustained campaign that lasted for several months. An important message of the conference was that PPE generally worked well for its designed purpose in the initial response. However, such equipment typically was not designed for the continuous use associated with a sustained response campaign. Firefighter turnout gear, for example, is constructed to be worn for, at most, hours. Accordingly, responders spoke of being hampered by basic problems such as wet garments and blistered feet.

Furthermore, at major terrorist-attack sites, emergency workers face a staggering range of hazards. Not only do they confront the usual hazards associated with building fires—flames, heat, combustion by-products, smoke—they also must be prepared to deal with rubble and debris, air choked with fine particles, human remains, hazardous materials (anhydrous ammonia, freon, battery acids), and the potential risk of secondary devices or a follow-on attack (see Figure 9-1). Conference participants indicated that many currently available PPE ensembles and training practices were not designed to protect responders from this range of hazards or were not supplied in sufficient quantity at the attack sites to meet the scale of the problem.

The scale of the terrorist events, their duration, and the range of hazards required that many emergency responders take on atypical tasks for which they were insufficiently equipped and trained. The nature of the destruction at the World Trade Center and the Pentagon reduced opportunities for primary reconnaissance and rescue—important tasks for fire fighters in large structural fires. Conversely, firefighters became engaged in activities they usually do not do: "busting up and hauling concrete," scrambling over a rubble pile, and removing victims and decayed bodies and body parts.

Construction workers were also deployed at the scenes and placed in hazardous environments early on. In all of the terrorist-incident responses, emergency medical personnel were on-scene, performing rescue operations, for example, in the rubble pile at the World Trade Center. Complicating activity at these already chaotic, hazardous, and demanding attack sites was the fact that the sites are also crime scenes. In addition, there were massive influxes of skilled and unskilled volunteers that created a significant challenge in managing the incident sites and assuring that all were properly protected.

In sum, the definition and roles of an *emergency responder* expanded greatly in the wake of the terrorist attacks, but few of the responders had adequate PPE, training or information for such circumstances.

PERSONAL PROTECTIVE EQUIPMENT PERFORMANCE AND AVAILABILITY

From the experiences at these attack sites, it is clear that there were significant shortfalls in the way responders were protected. Many responders suggested that the PPE even impeded their ability to accomplish their missions (see Figure 9-2).

Within the overall PPE ensemble used by responders at these sites, some equipment performed better than others. While head protection and high-visibility vests functioned relatively well for most responders, protective clothing and respirators exhibited serious shortcoming. Conference participants reported that the available garments did not provide sufficient protection against biological and infectious disease hazards, the heat of fires at the sites, and the demanding physical environment of unstable rubble piles, nor were they light and flexible enough to allow workers to move debris and enter confined spaces. Attendees also indicated that the available eye

FIGURE 9-1   Fire Fighter at World Trade Center Site
Source: FEMA

protection, while protecting well against direct impact injury, provided almost no protection against the persistent dust at the World Trade Center site.

Of all personal protective equipment, respiratory protection elicited the most extended discussion across all of the professional panels. Attendees indicated that under most circumstances, the self-contained breathing apparatus (SCBA) was grossly limited by both the weight of the systems and the short lengths of time (about 15 to 30 minutes) they can be used before their air bottles must be refilled. Most participants complained that respirators reduced their field of vision at best, and their facepieces fogged up at worst. Filters for air-purifying respirators (APRs) often did not match available facepieces, and many responders questioned the level of protection they provided, especially during anthrax responses.

FIGURE 9-2   Some Fire Fighters at World Trade Center Site Wearing PPE
Source: FEMA

For almost all protective technologies, responders indicated serious problems with equipment not being comfortable enough to allow extended wear during demanding physical labor. It was frequently observed that current technologies require a tradeoff between the amount of protection they provide and the extent to which they are light enough, practical enough, and wearable enough to allow responders to do their jobs. While conference attendees were concerned about having adequate protection, many were even more concerned about equipment hindering them from accomplishing their rescue and recovery missions in an arduous and sustained campaign. Respirators available at the sites were uncomfortable, causing many wearers to use them only intermittently (one participant dubbed them "neck protectors") or to discard them after a short period (see Figure 9-3).

For many fire fighters at the conference, PPE availability was as important a concern as PPE performance. Some health-and-safety panelists expressed a similar view. There was an acute shortage of respirators early in the response at the World Trade Center, for example. Subsequently, providing appropriate equipment to the large numbers of workers at these sites was made even more difficult because of the many types and brands of equipment that were being used by the various responder organizations or were being supplied from various sources. The problem was further exacerbated by a lack of interoperability among different types of equipment. These issues, coupled with the very large volume of equipment sent to the World Trade Center site, in particular, made it very difficult to match responders with appropriate equipment and supplies.

FIGURE 9-3 Fire Fighters at World Trade Center Site Wearing Respirators While Sifting Through Rubble
Source: FEMA

PERSONAL PROTECTIVE EQUIPMENT TRAINING AND INFORMATION

The responses to the terrorist attacks uncovered a range of PPE training and information needs. Before an incident occurs, those who are likely to be involved in a response should be trained on the proper selection and operation of personal protective equipment. Emergency medical technicians who were themselves treating casualties in the heart of the disaster site should have been wearing PPE but frequently were not, in large part because this equipment was not part of their standard training regimen.

The experiences in these incidents also showed that there is a need for significant on-site training to protect the health and safety of workers. The attack sites involved large numbers of workers, particularly construction workers and volunteers, many of whom were not familiar with most PPE. They needed to be trained in the proper selection and fitting of respirators, how to maintain them, and when to change filters. The situation with anthrax was more severe. Health and safety panel members felt that training support during the anthrax attacks was inadequate on all fronts: The response protocols were being developed *during* the actual response.

Emergency responders repeatedly stressed the importance of having timely and reliable health and safety information. "What kills rescue responders is the unknown," commented an emergency medical services (EMS) panel member. Several shortcomings were noted by conference participants. Special-operations and law-enforcement responders reported problems caused by different information sources telling them different things. Such information conflicts were often attributed to differences in risk assessment and PPE standards among reporting parties. Especially in the case of anthrax incidents, keeping up with changing information being provided by numerous agencies was a serious challenge for front-line responder organizations. For many conference participants, the problem was not a lack of information on hazards. Rather, they spoke of difficulties trying to manage and make sense of a surplus of information. Finally, conference attendees suggested that better and more consistent information provision could motivate responders to wear PPE and could decrease the tendency to modify it or take it off when it becomes uncomfortable.

SITE MANAGEMENT

One message that emerged clearly from virtually all panel discussions is that proper site management had a decisive effect on whether personal protective equipment was available, appropriately prescribed, used, and maintained.

The most critical need for site management is a coherent command authority. An effective command structure is essential to begin solving three critical issues affecting PPE: information provision, equipment logistics, and enforcement. Due to logistical problems early in the response, for example, supplies of PPE were misplaced, the stocks of equipment that were available were largely unknown, and responders often did not receive or could not find the equipment they needed.

Conference attendees also emphasized the need for immediate and effective perimeter or scene control. Initially, this entailed responders personally "holding people back" and isolating the scene. As the response evolved, it was necessary to erect a "hard perimeter," such as a chain link fence to make sure only essential personnel operating under the direction of the scene commander were on-site.

Conference attendees also indicated that enforcement of PPE use is very important. Although panelists acknowledged that there is a period early in a chaotic response when it is not practical to rigorously enforce the use of protective equipment, they indicated that strict enforcement must eventually begin in order to protect the health of the responders. Other factors that complicated enforcement of PPE use were the large number of organizations (with different PPE standards) operating on-site, the lack of a unified command, and shortcomings in scene control. Because of the difficulty of defining when it is appropriate to begin enforcing PPE use—and removing workers from the site if they do not comply with use requirements—attendees indicated that this role might be best played by an organization not directly involved in or affected by the incident.

RECOMMENDATIONS

After having discussed PPE performance, information and training, and site-management issues, NIOSH/RAND conference participants were asked to put forward concrete recommendations about technologies and procedures that could help protect the health and safety of emergency workers as they respond to acts of terrorism. The following points represent a brief sample of the themes that emerged and the solutions put forth by conference discussions.

Personal Protective Equipment Performance

- Develop guidelines for the appropriate PPE ensembles for long-duration disaster responses involving rubble, human remains, and a range of respiratory threats. If appropriate equipment is not currently available, address any roadblocks to its development. Such equipment could be applicable to other major disasters, such as earthquakes or tornadoes, as well as to terrorist attacks.

- Define the appropriate ensembles of PPE needed to safely and efficiently respond to biological incidents, threats, and false alarms. Key considerations include providing comparable levels of protection for all responders and addressing the logistical and decontamination issues associated with large numbers of responders in short time periods.

Personal Protective Equipment Availability

- Explore mechanisms to effectively outfit all responders at large incident sites with appropriate personal protective equipment as rapidly as possible.

- Examine any barriers to equipment standardization or interoperability among emergency-responder organizations. Strategies could include coordination of equipment procurement among organizations or work with equipment manufacturers to promote broader interoperability within classes of equipment.

Training and Information

- Define mechanisms to rapidly and effectively provide responders at incident sites with useful information about the hazards they face and the equipment they need for protection. Approaches could include more-effective coordination among relevant organizations and development of technologies that provide responders with individual, real-time information about their enforcement.

- Explore ways to ensure that responders at large-scale disaster sites are appropriately trained to use the protective equipment they are provided. All types of responders must be addressed, and mechanisms that provide training and experience with the equipment before a disaster occurs should be investigated.

- Consider logistical requirements of extended response activities during disaster drills and training. Such activities provide response commanders with information on the logistical constraints that could restrict response capabilities.

Management

- Provide guidelines and define organizational responsibilities for enforcing protective-equipment use at major disaster sites. While such guidelines must address the risks responders are willing to take when the potential exists to save lives, they must also consider that during long-term responses, the health and safety of responders should be a principal concern.

- Develop mechanisms to allow rapid and efficient control at disaster sites as early as possible during a response.

CONCLUDING REMARKS

The emergency workers and managers who attended the NIOSH/RAND conference provided a wealth of information on availability, use, performance, and management of personal protective equipment. Throughout the conference, a number of important issues were explicitly addressed during the meeting; others were implicit consequences of the lessons learned. This concluding chapter draws out several of these strategic policy issues for further reflection.

Guidelines

One of the clear messages of the conference was that most emergency workers do not believe that they are prepared with the necessary information, training, and equipment to cope with many of the challenges associated with the response to a major disaster such as the World Trade Center attack or for threats associated with anthrax and similar agents. These challenges include the large scale of the operations, the long duration of the response, the broad range of known and potential hazards encountered, and the assumption of nonstandard tasks by emergency responders.

Lessons learned from the response to the terrorist attacks suggest that near-term efforts to develop and upgrade equipment and operating guidelines could significantly improve the safety of emergency workers.

- Guidelines are needed for designing personal protective equipment ensembles appropriate for long-term responses to a range of major disasters.[1] An obvious case would be a disaster involving the collapse of one or more large buildings and the consequent need to work on rubble in the presence of a variety of hazards, including human remains, smoldering fires, and airborne contaminants derived from the building and its collapse.

- Recognizing that different responders have different personal protection requirements, these guidelines could also address the various professional groups working at a disaster site. Moreover, the guidelines should take into account the reality that individual responders may fulfill various tasks entailing different hazards and that hazards vary within the inner and outer perimeters of a disaster site.

- Protective equipment and safety guidelines could lead to better responses to biological incidents, not only for anthrax but for other potential biological threats.

- Well-designed guidelines and protocols could significantly improve real-time on-site hazard assessments. Essential elements include sensing equipment, measurement sites, organizational responsibilities and authorities, and data interpretation consistent with operational requirements.

- Discussions about the management of the terrorist attack sites often touched (sometimes indirectly) on sensitive and debated topics such as the appropriate time to declare an end to rescue efforts, the way off-duty and volunteer assistance should be managed, and the accommodation of VIPs and other concerned parties. Given the understandable difficulty of making such decisions in the midst of a response effort, site commanders could greatly benefit from guidelines developed in advance of an incident.

- To be useful, guidelines must be practical in the sense that they consider the capabilities of emergency-response organizations, are easy to use in the field, and do not unduly impair the ability of emergency responders to perform critical lifesaving missions.

Cost

The conference participants identified many new technologies for personal protection that would be desirable, based on the lessons learned from the terrorist attacks. Some argued that many desired technologies already exist and progress may simply be a

---

[1]By an ensemble, we mean the entire list of PPE responders should carry, including respirators, clothing, eye protection, sensors, etc.

question of procuring the appropriate equipment. Participants highlighted, however, that in the case of both existing and new technology, cost can be a very serious barrier to adoption of equipment by state and local response organizations. Powered-air respirators, for example, can cost ten times as much as the simpler nonpowered variety. Providing each emergency worker with his or her own ensemble of equipment specific to a range of hazards could be prohibitively expensive for most local emergency-response organizations.

Efforts could be directed toward making these technologies more affordable or, alternatively, developing efficient ways to deliver the appropriate equipment to incident sites. In instances where a desired technology is commercially available, expanding the number of prepositioned caches of such equipment that could be moved to response sites could be a good compromise solution. The know-how in supply logistics resident in the U.S. military could be helpful for developing supply strategies for the domestic emergency-response community. Another option would be preplanned equipment-sharing with non-neighboring emergency-response units.[2] For smaller departments, it may be appropriate to examine alternative approaches to increasing purchasing power, such as banding together and conducting coordinated procurements.

Research, Development, and Technology Transfer

Several panels put forth recommendations for new equipment and technologies, most of which were for modest and incremental improvements to existing technologies. Research and development (R&D), however, may yield significant benefits to the emergency-responder community. For example, a major theme that ran through many of the panels was the apparent tradeoff between the level of protection provided by equipment and the discomfort and physical burden the equipment placed upon those using it. Directing R&D toward advanced respirators, clothing sensors, and other safety gear may be able to reduce that tradeoff. Other areas suggested by the conference discussions include applications of information technology and communications systems for better management of worker safety at disaster sites and continued emphasis on technologies for locating responders buried or trapped under rubble.

As previously discussed, a theme that arose in several panel discussions was that the purchasing power of the emergency-response community was limited, given its relatively small size and tight budgets, especially at the local level. These factors constrain the community's ability to drive R&D on new technologies. However, much of the safety-related technology that is in use came through technology transfer from other industries, and in some cases, the military. Technology transfer is expected to continue to play an important role in providing emergency responders with improved safety equipment, for example, equipment using information technology, telecommunications, and advanced sensor systems originally developed for purposes other than emergency response.

Technology transfer can help reduce personal protective equipment costs by spreading R&D outlays across a larger user community. It can also speed the introduction of new technologies to the emergency-response community. But the emergency-response community also has special safety needs that may not be adequately met through technology transfer alone. Many at the meeting suggested that publicly supported R&D would be appropriate for addressing the safety needs of emergency responders. The recent terrorist attacks have raised awareness of this issue.

---

[2]In the event of a major disaster, neighboring emergency-response organizations are likely to be part of the response team and unavailable to share equipment.

Equipment Standardization and Interoperability

Equipment standardization and interoperability, as well as the development of more uniform training, maintenance, and use protocols, were mentioned as important needs throughout the conference discussions. Although these are not new issues, the scale and complexity of the terrorist attacks and the problems encountered in the responses appear to have drawn greater attention to them and have increased their importance as policy matters for all members of the emergency-response community. The recommendations put forth by conference participants indicate that these issues may be addressed from the top down (through promulgation of uniform safety standards) or from the bottom up (through greater interagency cooperation).

Safety Management

One of the most important lessons learned from the responders at the terrorist-attack sites is the importance of on-site safety management. Effective safety management is unlikely to be achieved if the overall site is not under a defined management structure, with clear lines of authority and responsibility. The operational side of safety management involved hazard monitoring and assessment, safety-equipment logistics and maintenance, site access control, health and safety monitoring, and medical treatment of emergency workers.

Given the magnitude of these tasks, conference participants argued that the safety officer at a disaster site should be an independent official whose sole responsibility is safety enforcement. In cases where incident sites are managed through a unified command structure, those responsible for responder safety could be part of that command.

From the federal perspective, an important issue is reassigning and clearly defining the roles and relationships of various federal agencies with health and safety responsibilities at a major disaster site.

Residential Fire, Keokuk, Iowa, December 22, 1999

Robert F. Duval

At approximately 8:24 A.M. on Wednesday, December 22, 1999, a fire was reported in a multi-family dwelling in Keokuk, Iowa. Several neighbors phoned the Keokuk 911 center to report that smoke was coming from a residence and that a woman was outside screaming that there were children trapped inside.

At the time the fire was reported, the on-duty force from the Keokuk Fire Department (an assistant chief, a lieutenant, and three fire fighters) was completing operations at a motor vehicle accident at a major intersection, two miles northwest of the fire scene. The dispatcher notified the units of the fire and the report of people trapped. Both units at the accident (Rescue 3 and Aerial 2) responded from the scene of the motor vehicle accident. During the response, additional calls were made to the 911 center, reporting heavy smoke coming from the house.

One member of the on-duty force of five fire fighters was committed in assisting the EMS crew on the ambulance and was en route to the Keokuk hospital at the time of the report of the house fire. The chief of the department became aware of the incident as he entered his office at the fire station. The chief responded from the fire station and went to the hospital to pick up the fire fighter that was with the ambulance crew.

Upon arrival at 8:28 A.M., the units found heavy smoke showing from a two-story multi-family dwelling on the northeast corner of a four-way intersection (see Figures 10-1 and 10-2). A water supply was established from a hydrant one-block southwest of the scene. Rescue 3 (R3), a 1500-gpm engine, laid a 5-inch in diameter supply line from the hydrant, while the lieutenant stayed at the hydrant to connect the line and activate the hydrant. Aerial 2 (A2), with a 50-foot (15.2 m) ladder and a 2000-gpm pump, continued to the scene.

The assistant chief requested six fire fighters be called back to duty as he arrived at the house in Aerial 2. As the two truck operators set up the apparatus, the assistant chief reportedly spoke to the female resident of the burning apartment. She reported that three of her children were still inside the apartment and that she tried but could not get them out. (She was able to exit the house via a second-floor window with her 4-year-old son, with the assistance of neighbors.) The assistant chief completed donning his protective clothing, including SCBA, and entered the right side apartment door.

Robert F. Duval is senior fire investigator and New England regional manager at the National Fire Protection Association.

Source: NFPA Fire Investigations Report Summary—Keokuk, Iowa, Residential Fire, 2000.

FIGURE 10-1   Fireground Activity Approximately 16 Minutes After Alarm
Source: Photo used with permission: C. Iutzi, *Daily Gate City Newspaper*

FIGURE 10-2   View of the Rear of the Fire Building Approximately 24 Minutes After Alarm
Source: Photo used with permission: C. Iutzi, *Daily Gate City Newspaper*

The chief arrived not long after the assistant chief entered the building. The chief ordered the two apparatus operators into the building to assist the assistant chief with the search for the children. Shortly thereafter, a fire fighter passed a 22-month-old male out the front door of the apartment to a police officer, who began CPR. The officer with the infant was then taken to a police car and transported to the hospital, six blocks west of the scene. A second child, an unresponsive 22-month-old female, was then passed out the door to the fire chief. With no EMS units yet on the scene, the chief chose to take the infant to the hospital in another police car, with a police captain driving. The fire chief conducted CPR on the infant during the one-minute ride to the hospital emergency room. He quickly handed the infant over to the emergency room staff and returned to the fire scene.

In the meantime, the fire fighter that arrived with the fire chief stretched a 1½-in. hoseline to the front door of the fire apartment and returned to don her SCBA. When the hoseline was charged, she noticed that the hoseline had burned through while at the entrance to the apartment. The fire fighter reported that the first level of the apartment was engulfed in flames, visible from her vantage point at Aerial 2 (see Figure 10-3).

FIGURE 10-3   Building Exterior Showing Fire Extension from Kitchen Upward to Children's Bedroom

The location and condition of the fire fighters and of the remaining child in the burning apartment was not known. The burned length of hose was removed, and the nozzle reconnected to the line as it was charged again. The fire fighter played a hose stream into the burning apartment. She was able to advance only 6 to 8 feet (1.8 to 2.4 m) into the apartment before being driven back by the intense heat.

The first two of the "call-back" fire fighters arrived in Engine 6 (E6–reserve unit). They were teamed with the lieutenant who was at the hydrant and had now walked the one block to the scene. The three were ordered to search the adjoining apartment for a resident that supposedly was still inside. The search was completed with nobody found. (The occupant was at a local restaurant.)

Efforts continued to contact the three fire fighters who were in the fire apartment. As additional call-back, fire fighters arrived in Aerial 1 ([A1] 100-foot [30.5 m] aerial unit with a 1500 gpm pump); they were ordered to begin to search for the missing fire fighters in the original fire apartment. As the fire was knocked back and a search could begin, fire fighters quickly found one fire fighter in the first floor room to the right of the main entrance corridor (see Figure 10-4). He had perished.

The assistant chief's body was found at the top of the stairs, not far from the body of the remaining child, a seven-year-old girl. The third fire fighter was found in the master bedroom to the right of the top of the stairs (see Figure 10-5). All had perished.

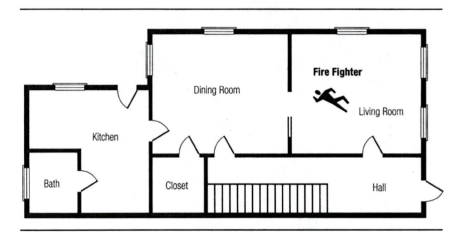

FIGURE 10-4   Location of Fire Fatality, First Floor

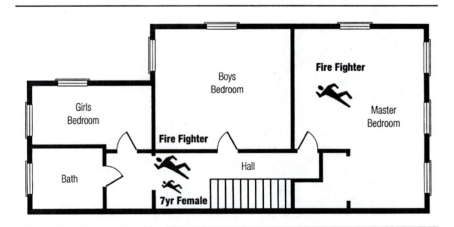

FIGURE 10-5   Location of Fire Fatalities, Second Floor

The remaining fire was extinguished at approximately 1:30 P.M. Overhaul was conducted until 3:30 P.M., and at that point units were placed back in service.

On the basis of the fire investigation and analysis, the NFPA has determined that the following significant factors may have contributed to the deaths of the three fire fighters:

- Lack of a proper building/incident size-up (risk-versus-benefit analysis)
- Lack of an established incident management system (IMS)
- Lack of an accountability system
- Insufficient resources (such as personnel and equipment) to mount interior fire suppression and rescue activities
- Absence of an established rapid intervention crew (RIC) and a lack of a standard operating procedure requiring a RIC

On the basis of the fire investigation and analysis, the NFPA has determined that the following significant factor may have contributed to the deaths of the three children:

- Lack of functioning smoke detectors within the apartment to provide early warning of a fire

11

Fire Fighter Fatalities, Marks, Mississippi, August 29, 1998

Robert F. Duval

At approximately 12:58 A.M. on Saturday, August 29, 1998, a fire was reported to the rear at the florist shop on Main Street in Marks, Mississippi. The fire reportedly began in a pile of cardboard and other combustible materials in the rear of the florist shop. The fire then spread through the open eaves of a storage building behind the florist shop. The 20-ft × 30-ft. (6.1-m × 9.1-m) storage building was used to store floral packing and display materials and also contained a 6-ft × 6-ft. (1.8-m × 1.8-m) cooler unit (see Figure 11-1). The building was connected to the main florist shop through a steel frame door. The florist shop was located in the middle of a block of buildings that contained a restaurant, a liquor store, a dry cleaner, and a lounge. The block of buildings was approximately 140 ft (42.6 m) in length and 60 ft (18.3 m) deep (see Figure 11-2).

Upon arrival of the first fire units, at 1:05 A.M., smoke and flame were showing from the eave line of the storage building. The fire department gained access to the storage building and began to extinguish the fire, within the building. An additional hoseline was deployed to protect a youth club building located 15 ft (4.6 m) south of the fire building. The Marks fire chief requested mutual aid from the Lambert Fire Department at 1:09 A.M.

With the fire in the storage building extinguished, salvage and overhaul was begun in the storage building and the adjoining florist shop. When the Marks fire chief entered the florist shop with the owner at about 1:25 A.M., he reported light smoke in the building. Further investigation revealed smoke showing from the attic space of the florist shop (see Figure 11-3). The chief then returned to the rear of the shop and ordered two Marks fire fighters to access the roof and check on conditions to determine if ventilation would be necessary.

The two Marks fire fighters placed a ground ladder at the rear of the liquor store and began to climb to the roof. One fire fighter was equipped with breathing apparatus, and the other was not. As they reached the roof, smoke conditions worsened, and

---

Robert F. Duval is senior fire investigator and New England regional manager at the National Fire Protection Association.

Source: NFPA Fire Investigations Report Summary—Marks, Mississippi, Fire Fighter Fatalities, 1999.

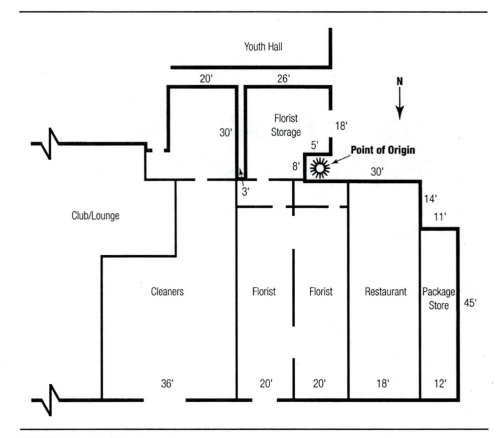

FIGURE 11-1   Building Layout

the fire fighter without breathing apparatus returned to the ground to find breathing apparatus to don. The fire fighter remaining on the roof then proceeded to walk over to the area of the florist shop. When he stepped from the roof of the restaurant onto the roof at the rear of the florist shop, at approximately 1:40 A.M., the weakened roof structure collapsed and he fell into the store, landing in the southeast storage room in the shop (see Figure 11-4). No one on the fire ground witnessed his falling through the roof. His location was unknown to the others on the fireground.

At the front of the florist shop, with smoke conditions worsening, a hoseline was stretched from the Lambert engine that had been positioned in the front of the restaurant. Two fire fighters (one from Marks and the other from Lambert) donned breathing apparatus and prepared to enter the front of the shop at about 1:55 A.M. (see Figure 11-5). The Marks fire fighter had also participated in the attack on the fire in the storage building and was on his third air cylinder. Within seconds of the two fire fighters' entry into the building, witnesses on the outside reported seeing the hoseline "jump." Immediately following this, the Lambert fire fighter stumbled out of the door and onto the sidewalk, stating that the fire fighter from Marks was still in the building. Fire fighters outside the shop, including the fire fighter who had just exited, entered the building and began searching for the Marks fire fighter lost at the front of the shop.

Numerous attempts were made to locate the fire fighter. Rescue efforts were hampered due to a lack of full air cylinders at the scene. A police officer had been dispatched to travel approximately 20 miles (32.2 km) to Batesville to refill the cylinders already depleted. The hose line that was used was located. The fire fighter, however,

FIGURE 11-2   View of Fire Buildings from Northeast Corner

FIGURE 11-3   View of Attic Space at Front (North) of Restaurant, with Blackened Bricks Indicating Attic Area

FIGURE 11-4   Area of Origin at Rear of Florist Shop, with Parapet Indicating Area Where Fire Fighter Fell Through Weakened Roof Structure

was not with the line. During the rescue attempts, the Marks fire chief was injured by broken glass in an effort to ventilate the florist shop.

Additional mutual aid was requested from the Batesville Fire Department at 2:03 A.M. Upon arrival of Batesville units at 2:25 A.M., fire fighters from Batesville began to assist in the search for the lost fire fighter in the front of the florist shop. The injured Marks fire chief turned command of the scene over to the Batesville chief while he sought medical attention for his injuries. At this point, additional mutual aid was requested from surrounding communities to assist in the search for the missing fire fighter and for help in the extinguishing of the fire.

Batesville fire fighters located the missing Marks fire fighter during the second search of the store, after 3:00 A.M. His body was found under a pile of debris within 24ft (7.3 m) of the front entrance.

FIGURE 11-5   Front of Florist Shop, with Attic Vents Showing Above Canopy

During the search efforts, the fire spread to the adjoining establishments. When the body of the fire fighter lost in the front of the florist shop was located and re-moved, the focus was again turned to extinguishment of the fire. At this point, it was determined that another fire fighter was missing, the Marks fire fighter who had gone to the roof in the rear of the block to ventilate. It was thought that he might be in the rear of the florist shop. Efforts were put forth to extinguish the fire in that and adjoin-ing areas so that another search effort could be mounted.

The fire in the rear of the block was under control at about 5:30 A.M., and the sec-ond missing fire fighter's body was found in a rear storage room of the florist shop around 6:00 A.M. (see Figure 11-6).

On the basis of the fire investigation and analysis, the NFPA has determined that the following significant factors directly contributed to the deaths of the two fire fighters:

- Lack of a fireground accountability system
- Ineffective use of an established incident management system (IMS)
- Failure to equip fire fighters with personal alert safety systems (PASS)

FIGURE 11-6   Rear Storage Room in Florist Shop Where Body of Second Fire Fighter was Found

- Lack of knowledge of the construction features of the building and how these features would affect the spread of fire in the concealed spaces, including the attic
- Insufficient resources (such as personnel and equipment, self-contained breathing apparatus [SCBA], and spare cylinders) to mount interior fire suppression and rescue activities.
- Absence of an established Rapid Intervention Crew (RIC) and the lack of a standard operating procedure requiring a RIC.

12

Six Career Fire Fighters Killed in Cold Storage and Warehouse Building Fire, Worcester, Massachusetts*

On December 3, 1999, six career fire fighters died after they became lost in a six-floor, mazelike, cold storage and warehouse building while searching for two homeless people and fire extension (see Figure 12-1). It is presumed that the homeless people had accidentally started the fire on the second floor sometime between 1630 and 1745 hours and then left the building (see Figure 12-2). An off-duty police officer who was driving by called central dispatch and reported that smoke was coming from the top of the building. When the first alarm was struck at 1815 hours, the fire had been in progress for about 30 to 90 minutes. Beginning with the first alarm, a total of five alarms were struck over a span of 1 hour and 13 minutes, with the fifth called in at 1928 hours. Responding were 16 apparatus, including 11 engines, 3 ladders, 1 rescue, and 1 aerial scope, and a total of 73 fire fighters. Two incident commanders (IC#1 and IC#2) in two separate cars also responded.

Fire fighters from the apparatus responding on the first alarm were ordered to search the building for homeless people and fire extension (see Figure 12-3). During the search efforts, two fire fighters (Victims 1 and 2) became lost, and at 1847 hours, one of them sounded an emergency message. A head count ordered by interior command confirmed which fire fighters were missing.

Fire fighters who had responded on the first and third alarms were then ordered to conduct search-and-rescue operations for Victims 1 and 2 and the homeless people. During these efforts, four more fire fighters became lost. Two fire fighters (Victims 3

---

*Editor's Note: This chapter consists of the Summary of the NIOSH Fatality Assessment and Control Evaluation Investigative Report #99F-47, September 27, 2000. The Fire Fighter Fatality Investigation and Prevention Program is conducted by the National Institute for Occupational Safety and Health (NIOSH). The purpose of the program is to determine factors that cause or contribute to fire fighter deaths suffered in the line of duty in order to develop strategies for preventing future similar incidents. For a copy of the full report or additional information, visit the program website at http://www.cdc.gov/niosh/firehome.html. For illustrative purposes two photos (Figures 12-4 and 12-5) were added from other sources and did not appear in the NIOSH report.

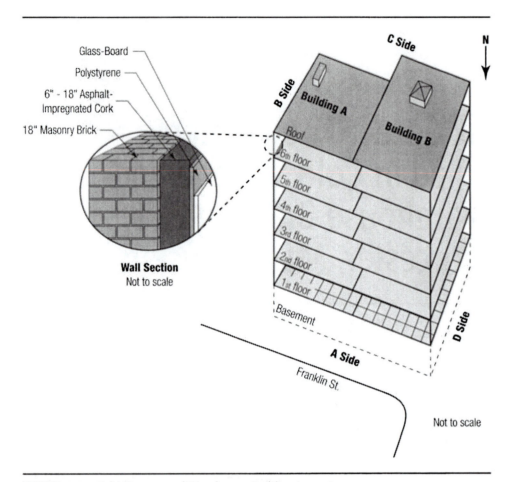

FIGURE 12-1   Cold-Storage and Warehouse Building Layout

and 4) became disoriented and could not locate their way out of the building. At 1910 hours, one of the fire fighters radioed command that they needed help finding their way out and that they were running out of air. Four minutes later he radioed again for help. Two other fire fighters (Victims 5 and 6) did not make initial contact with command nor anyone at the scene, and were not seen entering the building. However, according to the central dispatch transcripts, they may have joined Victims 3 and 4 on the fifth floor. At 1924 hours, IC#2 called for a head count and determined that six fire fighters were now missing.

At 1949 hours, the crew from Engine 8 radioed that they were on the fourth floor and that the structural integrity of the building had been compromised. At 1952 hours, a member from the Fire Investigations Unit reported to the chief that heavy fire had just vented through the roof on the C side. At 2000 hours, interior command ordered all companies out of the building, and a series of short horn blasts were sounded to signal the evacuation. Fire-fighting operations changed from an offensive attack, in-

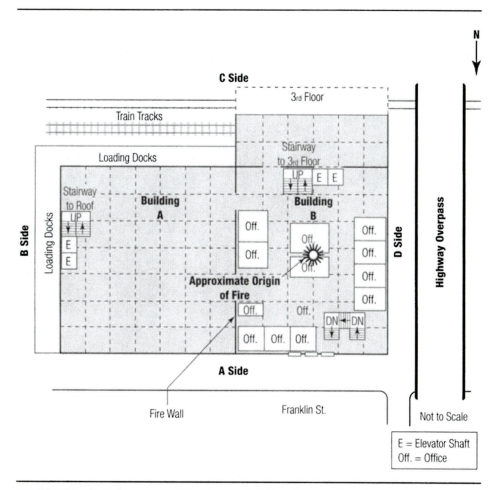

FIGURE 12-2 Cold Storage and Warehouse Building, Second Floor Layout, Showing Approximate Origin of Fire

cluding search and rescue, to a defensive attack with the use of heavy-stream appliances (see Figure 12-4).

After the fire had been knocked down, search-and-recovery operations commenced until recall of the box alarm 8 days later on December 11, 1999, at 2227 hours, when all six fire fighters' bodies had been recovered (see Figure 12-5). NIOSH investigators concluded that, to minimize the risk of similar occurrences, fire departments should

- Ensure that inspections of vacant buildings and pre-fire planning are conducted which cover all potential hazards, structural building materials (type and age), and renovations that may be encountered during a fire, so that the incident commander will have the necessary structural information to make informed decisions and implement an appropriate plan of attack

FIGURE 12-3   Fire-Fighting Efforts Shortly After First Alarm
Source: Photograph by Roger B. Conant

FIGURE 12-4   Water Being Poured on Lingering Fire
Source: AP/World Wide Photo/Paul Connors

FIGURE 12-5   Search-and-Recovery Operations Underway
Source: Worcester Telegram & Gazette/ Paula Perazzi Swift

- Ensure that the incident command system is fully implemented at the fire scene
- Ensure that a separate incident safety officer, independent from the incident commander, is appointed when activities, size of fire, or need occurs, such as during multiple alarm fires, or responds automatically to pre-designated fires
- Ensure that standard operating procedures (SOPs) and equipment are adequate and sufficient to support the volume of radio traffic at multiple-alarm fires
- Ensure that incident command always maintains close accountability for all personnel at the fire scene
- Use guide ropes/tag lines securely attached to permanent objects at entry portals and place high-intensity floodlights at entry portals to assist lost or disoriented fire fighters in emergency escape
- Ensure that a rapid intervention team is established and in position upon their arrival at the fire scene
- Implement an overall health and safety program such as the one recommended in NFPA 1500, *Standard on Fire Department Occupational Safety and Health Program*
- Consider using a marking system when conducting searches
- Identify dangerous vacant buildings by affixing warning placards to entrance doorways or other openings where fire fighters may enter
- Ensure that officers enforce and fire fighters follow the mandatory mask rule per administrative guidelines established by the department

- Explore the use of thermal imaging cameras to locate lost or downed fire fighters and civilians in fire environments

In addition, manufacturers and research organizations should conduct research into refining existing and developing new technology to track the movement of fire fighters on the fireground.

South Canyon Fire Investigation (Storm King Mountain)*

On July 2, 1994, during a year of drought and at a time of low humidity and record high temperatures, lightning ignited a fire 7 miles west of Glenwood Springs, Colorado (see Figure 13-1). The fire was reported to the Bureau of Land Management on July 3 as being in South Canyon, but later reports placed it near the base of Storm King Mountain. The fire began on a ridge, which was paralleled by two canyons or deep drainages, called in this report the east and west drainages (see Figures 13-2 and 13-3). In its early stages the fire burned in the pinyon-juniper fuel type and was thought to have little potential for spread.

Dry lightning storms had started 40 new fires in BLM's Grand Junction District in the 2 days before the South Canyon fire started, requiring the district to set priorities for initial attack. Highest priority was given to fires threatening life, residences, structures, utilities, and to fires with the greatest potential for spread. All initial attack fire-fighting resources on the Grand Junction District were committed to the highest priority fires. In response to a request from the Grand Junction District, the Garfield County Sheriff's Office and White River National Forest monitored the South Canyon fire.

Over the next 2 days the South Canyon fire increased in size, the public expressed more concern about it, and some initial attack resources were assigned. On the afternoon of July 4, the district sent two engines. Arriving at 6:30 P.M. at the base of the ridge near Interstate 70, the crew sized up the fire but decided to wait until morning to hike to the fire and begin fire-fighting efforts.

---

*Editor's Note: This chapter consists of the Executive Summary from the *Report of the South Canyon Fire Accident Investigation Team, August 17, 1994*, National Interagency Fire Center, Boise, Idaho. For illustrative purposes, photographs and line drawings have been added from Res. Pap. RMRS-RP-9: *Fire Behavior Associated with the 1994 South Canyon Fire on Storm King Mountain, Colorado*, by Bret W. Butler et al., Ogden, UT. U.S. Department of Agriculture, Forest Service, Rocky Mountain Research Station. These illustrations are not part of the original report. For the PDF version see http://www.fs.fed.us/rm/pubs/ rmrs_rp09.pdf. For additional information see *Final Report of the Interagency Management Review Team, South Canyon Fire, June 26, 1995*, at http://www.fs.fed.us/land/scanyon2.htm.

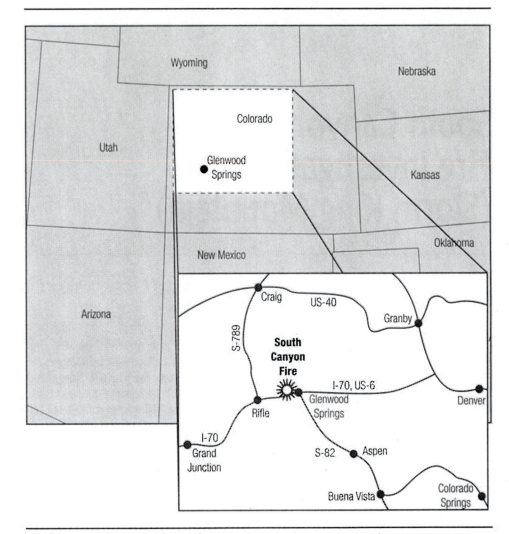

FIGURE 13-1　Overview of South Canyon Fire Location (not to scale)

　　The next morning, a seven-person BLM/Forest Service crew hiked $2\frac{1}{2}$ hours to the fire, cleared a helicopter landing area (Helispot 1) and started building a fireline on its southwest side. During the day an air tanker dropped retardant on the fire. In the evening the crew left the fire to repair their chainsaws. Shortly thereafter, eight smoke-jumpers parachuted to the fire and received instructions from the incident comman-der to continue constructing the fireline. The fire had crossed the original fireline, so they began a second fireline from Helispot 1 downhill on the east side of the ridge. After midnight they abandoned this work due to the darkness and the hazards of rolling rocks (see Figure 13-4).

　　On the morning of July 6, the BLM/Forest Service crew returned to the fire and worked with the smokejumpers to clear a second helicopter landing area (Helispot 2). Later that morning eight more smokejumpers parachuted to the fire and were assigned

FIGURE 13-2   Aerial View of South Canyon Fire Site Looking North up West Drainage
Toward Storm King Mountain
Source: J. Kautz, U.S. Forest Service, Missoula, MT

to build the fireline on the west flank. Later, ten Prineville Interagency Hotshot Crew
members arrived, and nine joined the smokejumpers in line construction. Upon arrival, the remaining members of the hotshot crew were sent to help reinforce the fireline on the ridgetop.

At 3:20 P.M. a dry cold front moved into the fire area. As winds and fire activity
increased, the fire made several rapid runs with 100-flame lengths within the existing
burn (see Figure 13-5).

FIGURE 13-3   Aerial Oblique View from Upper East Drainage Looking Southwest
Source: J. Kautz, U.S. Forest Service, Missoula, MT

At 4:00 P.M. the fire crossed the bottom of the west drainage and spread up the drainage on the west side. It soon spotted back across the drainage to the east side beneath the firefighters and moved onto steep slopes and into dense, highly flammable Gambel oak. Within seconds a wall of flame raced up the hill toward the fire fighters on the west flank fireline. Failing to outrun the flames, 12 fire fighters perished. Two helitack crew members on top of the ridge also died when they tried to outrun the fire to the northwest. The remaining 35 fire fighters survived by escaping out the east drainage or seeking a safety area and deploying their fire shelters.

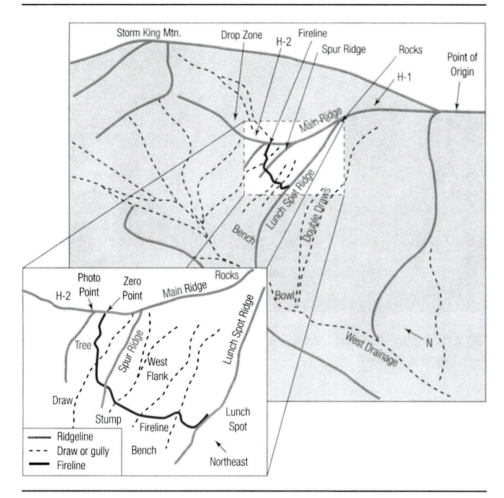

FIGURE 13-4   Oblique View of West-Facing Slope of Main Ridge Showing Fireline

THE INVESTIGATION

Within 3 hours of the blowup, an interagency team was forming to investigate the entrapment on the South Canyon fire. The team first met on the evening of July 7. Team members were given their assignments, and the team presented a charter to the chief of the USDA Forest Service and the director of the Bureau of Land Management. Les Rosenkrance, BLM's Arizona State Director, was designated team leader.

In the next few days the team investigated the fire and fatality sites and began a series of 70 interviews with witnesses. In addition, the team met once or twice a day to discuss progress, clarify assignments, plan their report, and review their findings. On July 22, with the interviews and much of the investigation report completed, the team adjourned. The following week some team members met in Phoenix, Arizona to complete work on the incident overview. On August 9–11, the team reconvened to review a draft of the completed report in preparation for its publication.

FIGURE 13-5   View Southwest Toward Fireline and Fire Burning Behind Spur Ridge
Source: Courtesy of S. Doehring

DIRECT CAUSES

The investigation team determined that the direct causes of the entrapment in the South Canyon fire are as follows.

Fire Behavior

Fuels.   Fuels were extremely dry and susceptible to rapid and explosive spread. The potential for extreme fire behavior and reburn in Gambel oak was not recognized on the South Canyon fire.

Weather.   A cold front, with winds of up to 45 mph, passed through the fire area on the afternoon of July 6.

Topography.   The steep topography, with slopes from 50 to 100 percent, magnified the fire behavior effects of fuel and weather.

Predicted Behavior.   The fire behavior on July 6 could have been predicted on the basis of fuels, weather, and topography, but fire behavior information was not requested or provided. Therefore critical information was not available for developing strategy and tactics.

Observed Behavior.   A major blowup did occur on July 6 beginning at 4:00 P.M. Maximum rates of spread of 18 mph and flames as high as 200 to 300 feet made escape by fire fighters extremely difficult.

Incident Management

Strategy and Tactics. Escape routes and safety zones were inadequate for the burning conditions that prevailed. The building of the west flank downhill fireline was hazardous. Most of the guidelines for reducing the hazards of downhill line construction in the *Fireline Handbook* were not followed. Strategy and tactics were not adjusted to compensate for observed and potential extreme fire behavior. Tactics were also not adjusted when Type 1 crews and air support did not arrive on time on July 5 and 6.

Safety Briefing and Major Concerns. Given the potential fire behavior,

- The escape route along the west flank fireline was too long and too steep.
- Eight of the 10 Standard Firefighting Orders were compromised.
- Twelve of the 18 Watch Out Situations were not recognized, or proper action was not taken.
- The Prineville Interagency Hotshot Crew (an out-of-state crew) was not briefed on local conditions, fuels, or fire weather forecasts before being sent to the South Canyon fire.

Involved Personnel Profile

The "can do" attitude of supervisors and fire fighters led to a compromising of standard fire-fighting orders and a lack of recognition of the 18 watch-out situations. Despite the fact that they recognized that the situation was dangerous, fire fighters who had concerns about building the west flank fireline questioned the strategy and tactics but chose to continue with line construction.

Equipment

- Personal protective equipment performed within design limitations, but wind turbulence and the intensity and rapid advance of the fire exceeded these limitations or prevented effective deployment of fire shelters.
- Packs with fusees taken into a fire shelter compromised the occupant's safety.
- Carrying tools and packs significantly slowed escape efforts.

CONTRIBUTORY CAUSES

The following factors contributed to the entrapment on the South Canyon fire.

Incident Management and Control Mechanisms

- The initial suppression action was delayed for 2 days because of higher priority fires on the Grand Junction District.
- Air support was inadequate for implementing strategies and tactics on July 6.

Support Structure

- The above-normal fire activity overtaxed a relatively small fire-fighting organization at the Grand Junction District and Western Slope Fire Coordination Center.
- Detailed fire weather and fire behavior information was not given to fire fighters on the South Canyon fire.

- Dispatching procedures and communications with the incident commander did not give a clear understanding of what resources (crews and air support) would be provided to the fire in response to requests and orders.

- Unclear operating procedures between the Western Slope Fire Coordination Center and the Grand Junction Districts fire organizations resulted in confusion about priority setting, operating procedures, and availability of fire-fighting resources, including initial attack resources (i.e., helitack fire fighters, smokejumpers, and retardant aircraft). The lack of definition limited the effectiveness in the timing and priority of the suppression of the South Canyon fire.

- The lack of Grand Junction District and Colorado State Office management oversight, technical guidance, and direction resulted in uncertainty concerning the roles and responsibilities of the Western Slope Fire Coordination Center and the Grand Junction District.

Checklists, Forms, and Other Resource Materials

As the current writer discovered when serving as a member of the NFPA Fire Service Occupational Safety and Health Technical Committee, significant resources exist to assist the user and enforcer of NFPA 1500, *Standard on Fire Department Occupational Safety and Health Program*. The 1992 edition of NFPA 1500 provided the user with a checklist on the implementation of the standard. The next two editions, 1997 and 2002, have provided updates to that checklist.

The chapters in Part IV provide checklists, forms, and other resource materials for readers to use within their jurisdictions. To assist the fire service, many progressive fire departments have contributed the forms and checklists their members use in the field.

Chapter 14, "NFPA Forms, Worksheets, and Checklists," provides four reproducible safety and health-related documents that have appeared in NFPA documents. Chapter 15, "Line-of-Duty Death Resources for Fire Departments," provides a wealth of information from the National Fallen Fire Fighters Foundation, including their Line-of-Duty Death Action Checklist.

Chapter 16, "Sample Incident Management System Forms," includes twelve forms on incident management systems—nine from the National Inter-Agency Incident Management System and three from the Phoenix Fire Department. Chapter 17, "Health, Fitness, and Wellness Resource Directory," provides updated information, including website addresses, for organizations concerned with health, fitness, and wellness.

Chapter 18, "Guide for Fire Department Administrators," which appears as an annex to the 2003 edition of NFPA 1582, *Standard on Medical Requirements for Fire Fighters and Information for Fire Department Physicians*, explores legal considerations in applying the standard. Chapter 19, "Fire Service Joint Labor Management Wellness-Fitness Initiative," this book's last chapter, is the introductory chapter from the manual of the Fire Service Joint Labor Management Wellness-Fitness Initiative and provides an overview of the fitness and wellness program.

NFPA Forms, Worksheets, and Checklists

T his chapter contains full-size reproducible copies of forms that are included in several of the fire service occupational safety codes and standards. The forms provide users with a template for achieving compliance with the standards. Users may alter the forms to fit the needs of their own departments. The included forms are

- Form 14-1 Sample Fire Department Facilities Safety Checklist
- Form 14-2 Sample Fire Protection District—Station Inspection Form
- Form 14-3 Sample Fire Department Occupational Safety and Health Program Worksheet
- Form 14-4 Sample Fire Department Infectious Exposure Form

**[ANYTOWN] FIRE DEPARTMENT
FACILITIES SAFETY CHECKLIST**

I. General

_____ The required Occupational Safety and Health workplace poster shall be displayed in the station, as required, where all employees are likely to see it.

_____ Emergency instructions and telephone numbers shall be available for the general public in the event of an emergency and fire personnel are out of quarters.

II. Housekeeping

_____ All rooms, offices, hallways, storage rooms, and the apparatus floor shall be kept clean and orderly and in a sanitary condition.

_____ All hallways and/or passageways shall be free from any type of protruding objects such as nails, splinters, and holes.

_____ All waste containers shall be emptied regularly.

_____ Waste containers shall be provided in the kitchen and/or eating areas. These containers shall have tight lids.

_____ All areas of the station shall be adequately illuminated.

_____ Stairways shall be in good condition with standard railings provided for every flight having four or more risers.

_____ Portable ladders shall be adequate for their purpose, in good condition, and have secure footing.

_____ Fixed ladders shall be equipped with side rails, cages, or special climbing devices.

_____ Smoking shall not be permitted in designated no smoking areas.

_____ Containers of all cleaning agents shall be carefully labeled per the 1910.1200 VOSH standard.

_____ First-aid supplies shall be available and clearly identified as to location.

_____ Shower curtains should provide adequate protection to prevent floors from becoming excessively wet and slippery.

_____ Cooking appliances and eating utensils should be kept clean and in good working order.

III. Exits

_____ All exits shall be visible and unobstructed.

_____ All exits shall be marked with a readily visible sign that is illuminated.

_____ Doors that might be mistaken for exits shall be marked "Not an Exit."

_____ Exits and exit signs shall be free of decorations, draperies, and/or furnishings.

_____ Primary exit routes shall be obvious, marked, and free of obstructions.

_____ Exits should be wide enough for easy access.

IV. Walking and Working Surfaces

_____ Floors shall be kept as clean and dry as possible.

_____ Adequate lighting shall be provided in all working areas.

_____ Fire fighters' routes to slide poles or to apparatus shall be completely free of projections, tripping hazards, loose objects, or other impediments.

_____ Beds shall be located so as to cause minimum interference during turnout of fire fighters.

_____ Handrails shall be of sufficient strength and proper design for all stairways and floor openings.

_____ All slide pole floor openings shall be provided with safety enclosures.

_____ A safety mat shall be positioned at the bottom of the slide pole.

_____ The slide pole shall be regularly inspected and maintained.

V. Apparatus Floor and Maintenance Areas

_____ Ladders, pike poles, and other items projecting from the apparatus shall be clearly marked with bright colored flags, stripes, or other identification to warn against "headbump" accidents.

_____ Apparatus overhead doors shall be maintained in safe, operating condition.

_____ Apparatus doors shall have adequate space for proper clearance for vehicles.

_____ Maintenance pits shall be adequately covered, sufficiently lighted, and ventilated.

_____ Pit boundaries shall be clearly marked.

_____ The pit floor shall be kept clean and as dry as possible.

_____ Fire fighters shall use adequate eye protection when working with grinders, drills, saws, welding equipment, and other tools likely to present an eye hazard.

_____ Eye protection shall be worn by personnel when working under vehicles.

_____ Eye protection shall be used and be in good condition.

(Page 1 of 2)

FORM 14-1 Sample Fire Department Facilities Safety Checklist

Source: NFPA 1500, _Standard on Fire Department Occupational Safety and Health Program_, 2002 ed., Figure G.1

_____ Work rests on grinders shall be adjusted to within ¹/₈ in. (.32 cm) to the grinding wheel.

_____ Grinders and grinding wheels shall be adequately guarded. The safety guard shall cover the spindle end, nut, and the flange projections.

_____ All power tools shall be provided with proper guarding for electrical, cutting, and moving parts.

_____ Maintenance hand tools shall be safely stored when not being used. They shall be inspected periodically and maintained to ensure their safe condition.

Conditions to check for safety are as follows:

_____ The tools shall be clean.

_____ The handles/grips shall not be broken.

_____ There shall be no worn, defective points/parts on the tool.

_____ There shall be no parts missing.

_____ Pulleys and belts shall be properly guarded.

_____ Chain drives and sprockets shall be guarded.

_____ Air cleaning nozzles shall not emit more than 30 psi deadened pressure. This information will be stamped on the nozzle.

_____ A spotter shall be used when vehicles are backed up, especially as a vehicle is driven over a pit.

VI. Fire Prevention and Protection

_____ Portable fire extinguishers shall be maintained in a fully operable condition and kept in designated places when not in use. They shall be inspected on a monthly basis.

_____ Fire extinguishers shall be of the proper type for the expected hazards.

_____ The fire extinguisher shall have a durable tag securely attached to show the maintenance or recharge date. Also, the initials or signature of the person who performed the inspection shall be on the tag.

_____ The fire alarm system shall be tested on a quarterly basis, if the station is so equipped.

_____ If the station is so equipped, the sprinkler system shall be serviced by a qualified person.

_____ The minimum amount of clearance, 18 in. (45.7 cm), shall be maintained below the sprinkler heads.

_____ Smoke detectors, which are in stations not equipped with a fire alarm system, shall be tested the first Tuesday of each month.

VII. Hazardous Materials

_____ Cylinders of compressed gases shall be stored in an upright position away from combustible materials.

_____ Flammable and combustible materials shall be stored in tanks or closed containers per NFPA 30, _Flammable and Combustible Liquids Code._

_____ Safety containers with self-closing lids shall be used for the storage of flammable liquids and soiled, oily rags.

_____ Gasoline and diesel pumps shall be checked on a weekly basis for proper working order and the condition of the nozzles and hoses.

VIII. Electrical Wiring, Fixtures, and Controls

_____ Electrical cords shall be strung so that they do not hang on pipes, nail hooks, and so forth.

_____ Conduit shall be attached to all supports and tightly connected to junction and outlet boxes.

_____ All electrical cords shall be checked for fraying.

_____ All equipment shall be securely mounted to the surface on which it sits.

_____ Flexible cords and cables shall not be used as a substitute for fixed wiring.

_____ All extension cords shall be properly grounded and approved.

_____ All electrical tools, whether department owned or personnel property, shall be properly protected for damaged power cords, plugs, worn switches, defective ground circuits, or other faults that could render them unsafe for use.

_____ Electrical switches and circuit breakers shall be marked to show their purpose.

IX. Other

_____ Portable heaters used in stations shall be placed out of travel routes and away from combustibles, and, if turned over, they shall turn themselves off.

_____ Any situations that warrant a concern shall be brought to the attention of the health and safety officer.

X. Comments/Explanations

_____

_____

_____

_____

_____

_____

_____

_____

_____

(Page 2 of 2)

FORM 14-1   _(Continued)_

[ANYTOWN] FIRE PROTECTION DISTRICT — STATION INSPECTION FORM

Station: _____   Date: _____

Shift: _____   Inspected By: _____

Officer responsible for corrections: _____

Answer all questions with yes or no. Explain any no answers. Comment on the bottom of the page.

General Work Environment

_____ Are all work sites clean and orderly?

_____ Are all work surfaces kept dry or are appropriate
means taken to assure the surfaces are slip-resistant?

_____ Are all combustibles stored properly?

_____ Are all bathroom facilities clean and sanitary?

_____ Is the kitchen clean and sanitary?

_____ Is the day room clean?

_____ Are the sleeping quarters clean?

_____ Are there proper labels on all containers?

_____ Are the apparatus room and shop area clean?

_____ Is the outside of the station clean and cared for?

_____ Are station log and all computer reports
(INFERS, Training) up to date and correct?

Comments: _____

_____

_____

_____

_____

Environmental Controls

_____ Are all electrical fixtures working?

_____ Is the furnace working properly?

_____ Are the furnace filters clean?

_____ Are there any combustibles around the furnace
or hot water heater?

_____ Is the hood over the range clean?

_____ Is the apparatus room exhaust system in use
and working properly?

_____ Is the air-conditioning or evaporative cooler clean
and working properly?

_____ Are all station exhaust fans working?

_____ Are floor drains clean and draining properly?

_____ Are all fire extinguishers up to date?

_____ Are smoke detectors working?

_____ Are material safety data sheets up-to-date?

_____ Are extension cords used as permanent wiring?

_____ Does all personnel have uniforms that meet the requirement of NFPA 1975, *Standard on Station / Work
Uniforms for Fire and Emergency Services*?

_____ Are station CO detectors in place and working
properly?

Comments: _____

_____

_____

_____

_____

Apparatus Number _____

_____ Are all apparatus tires safe?

_____ Is there any broken or defective glass in any
apparatus window?

_____ Are all lights on apparatus working (i.e., red lights,
working lights, headlights, taillights, marker lights,
and so forth)?

_____ Is apparatus clean?

_____ Is all equipment on apparatus and working properly?

_____ Are all apparatus checks and paperwork up-to-date
and filled out properly? (e.g., SCBA check-off sheets)

_____ Is apparatus ready to respond? (ask officer)

_____ Is medical equipment clean and inspected?

_____ Are all seat belts in place and working?

_____ Are all safety gates in place and working?

_____ Are all intercom headsets in place and working?

_____ Are hand-held radios in place and working?

_____ Is mobile radio working?

_____ Are map books on apparatus and up-to-date?

_____ Are building surveys on apparatus and up-to-date?

Comments: _____

_____

_____

_____

_____

(Page 1 of 1)

FORM 14-2   Sample Fire Protection District–Station Inspection Form

Source: NFPA 1500, *Standard on Fire Department Occupational Safety and Health Program*, 2002 ed., Figure G.1

NFPA 1500
FIRE DEPARTMENT OCCUPATIONAL SAFETY AND HEALTH PROGRAM WORKSHEET

Fire Department: _____   Date: _____

Person(s) Completing Worksheet

Name: _____   Title: _____

Name: _____   Title: _____

Name: _____   Title: _____

Standard Content	New in 2002 Ed.	Compliance	Partial Compliance	Compliance with Admin. Action	Expected Compliance Date	Compliance with Fiscal Action	Estimated $$	Targeted Compliance Date	Remarks or Modifications
Chapter 1 Administration									
1.4 Equivalency									
1.4.1 Equivalency levels of qualifications									
1.4.2 Training, education, competency, safety									
Chapter 4 Organization									
4.1 Fire Dept. Organizational Statement									
4.1.1 Written statement or policy									
4.1.2 Operational response criteria									
4.1.3 Statement available for inspection									
4.2 Risk Management Plan									
4.2.1 Written risk management plan									
4.2.2 Risk management plan coverage									
4.2.3 Risk management plan components									
4.3 Safety and Health Policy									
4.3.1 Written fire dept. occupational safety and health policy									
4.3.2 Program complies with NFPA 1500									
4.3.3 Evaluate effectiveness of plan									
4.3.3.1 Occupational safety and health program audit									
4.4 Roles and Responsibilities									
4.4.1 Fire dept. responsibility									
4.4.2 Comply with laws									
4.4.3 Fire dept. rules, regulations, and SOPs									

Copyright © 2002 National Fire Protection Association   (Page 1 of 25)

FORM 14-3   Sample Fire Department Occupational Safety and Health Program Worksheet

Source: NFPA 1500, *Standard on Fire Department Occupational Safety and Health Program*, 2002 ed., Figure B.1

NFPA 1500
FIRE DEPARTMENT OCCUPATIONAL SAFETY AND HEALTH PROGRAM WORKSHEET

Standard Content	New in 2002 Ed.	Compliance	Partial Compliance	Compliance with Admin. Action	Expected Compliance Date	Compliance Date	Compliance with Fiscal Action	Estimated $$	Targeted Compliance Date	Remarks or Modifications
4.4.4 Accident investigation procedure										
4.4.5 Members										
4.4.6 Fire dept. vehicles, equipment, facilities										
4.4.7 Corrective action to avoid repetitive occurrences										
4.4.8 Accident investigation records										
4.4.9 Individuals shall cooperate, participate, and comply										
4.4.10 Right to be protected and participate										
4.4.11 Member organization shall cooperate										
4.4.12 Collective rights										
4.5 Occupational Safety and Health Committee										
4.5.1 Establish committee										
4.5.1.1 Committee make-up										
4.5.1.2 Permit to include other persons										
4.5.2 Committee study and review										
4.5.3 Regular meetings										
4.5.3.1 Meetings at least every 6 months										
4.5.3.2 Minutes retained and available to members										
4.6 Records										
4.6.1 Accidents, injury, illness, exposures, death										
4.6.2 Occupational exposures										
4.6.3 Health (confidential)										
4.6.4 Training										
4.6.5 Vehicles and equipment										
4.7 Functions of the Health and Safety Officer										
4.7.1 NFPA 1521, fire dept. health and safety officer	✓									
4.7.1.1 Complies with NFPA 1021	✓									

Copyright © 2002 National Fire Protection Association   (Page 2 of 25)

FORM 14-3 *(Continued)*

NFPA 1500
FIRE DEPARTMENT OCCUPATIONAL SAFETY AND HEALTH PROGRAM WORKSHEET

Standard Content	New in 2002 Ed.	Compliance	Partial Compliance	Compliance with Admin. Action	Expected Compliance Date	Compliance with Fiscal Action	Estimated $$	Targeted Compliance Date	Remarks or Modifications
4.7.2 Fire dept. health and safety officer manage program									
4.7.3 Communicate plan to members									
4.7.4 Written risk plan available to members									
4.7.5 Monitor effectiveness of plan, revise annually	✓								
4.7.6 Develop plan into the incident management system	✓								
4.8 Laws, Codes, and Standards									
4.8.1 Develop, review, revise SOPs	✓								
4.8.1.1 Ensure compliance	✓								
4.8.1.2 Submit recommendations	✓								
4.8.2 Report adequacy	✓								
4.8.3 Define role of health and safety officer	✓								
4.9 Training and Education									
4.9.1 Provide safety and health training	✓								
4.9.2 Training arising from investigations	✓								
4.9.3 Safety supervision at training	✓								
4.9.4 NFPA 1403, live fire training									
4.9.5 Pre-burn inspection									
4.9.6 Distribute health and safety information									
4.10 Accident Prevention									
4.10.1 Health and safety officer manage accident prevention program	✓								
4.10.2 Health and safety officer delegate the development, participate, review, or supervise the program	✓								
4.10.3 Instruction for safe work practices	✓								
4.10.4 Addresses training and testing of all fire dept. drivers	✓								
4.10.5 Survey operations, procedures, equipment, and fire dept. facilities	✓								
4.10.6 Report recommendations	✓								

(Page 3 of 25)

FORM 14-3   *(Continued)*

NFPA 1500
FIRE DEPARTMENT OCCUPATIONAL SAFETY AND HEALTH PROGRAM WORKSHEET

Standard Content	New in 2002 Ed.	Compliance	Partial Compliance	Compliance with Admin. Action	Expected Compliance Date	Compliance with Fiscal Action	Estimated $$	Targeted Compliance Date	Remarks or Modifications
4.11 Accident Investigation, Procedures, and Review									
4.11.1 Develop and implement procedures									
4.11.1.1 Occupational injuries and illnesses are treated									
4.11.2 Occupational injuries, illnesses, exposures, and fatalities, or other potentially hazardous conditions									
4.11.3 Develop corrective recommendations									
4.11.4 The health and safety officer submit such recommendations									
4.11.5 The health and safety officer develop and review procedures									
4.11.5.1 Comply with local, state, and federal requirements									
4.11.6 Review the procedures employed during any unusually hazardous operation									
4.12 Records Management and Data Analysis									
4.12.1 Maintain records of all accidents, occupational deaths, injuries, illnesses, and exposures									
4.12.2 The health and safety officer manage information									
4.12.3 The health and safety officer identify and analyze safety and health hazards and develop corrective actions	✓								
4.12.4 The health and safety officer ensure that records are maintained as specified	✓								
4.12.5 The health and safety officer maintain records of recommendations	✓								
4.12.6 Maintain records of all measures taken to implement safety and health procedures and accident prevention methods	✓								
	✓								

(Page 4 of 25)

FORM 14-3 *(Continued)*

NFPA 1500
FIRE DEPARTMENT OCCUPATIONAL SAFETY AND HEALTH PROGRAM WORKSHEET

Standard Content	New in 2002 Ed.	Compliance	Partial Compliance	Compliance with Admin. Action	Expected Compliance Date	Compliance Date Fiscal Action	Compliance with Estimated $$	Targeted Compliance Date	Remarks or Modifications
4.13 Apparatus and Equipment									
4.13.1 Review specifications for new apparatus, equipment, protective clothing, and protective equipment	✓								
4.13.2 Assist and make recommendations regarding the evaluation of new equipment	✓								
4.13.3 Assist and make recommendations regarding the service testing	✓								
4.13.4 Develop, implement, and maintain a protective clothing and protective equipment program	✓								
4.14 Facility Inspection									
4.14.1 Inspect fire dept. facilities	✓								
4.14.2 Safety or health hazards or code violations are corrected	✓								
4.15 Health Maintenance									
4.15.1 Complies with Chapter 8	✓								
4.15.2 Health and wellness programs	✓								
4.16 Liaison									
4.16.1 The health and safety officer is member of the fire dept. occupational safety and health committee									
4.16.2 The health and safety officer reports the recommendations of the fire dept. occupational safety and health committee									
4.16.3 The health and safety officer submits recommendations on occupational safety and health	✓								
4.16.4 The health and safety officer provides information and assistance	✓								
4.16.5 The health and safety officer maintains a liaison with staff officers	✓								
4.16.6 The health and safety officer maintains a liaison with outside agencies	✓								
4.16.7 The health and safety officer maintains a liaison with the fire department physician	✓								

(Page 5 of 25)

FORM 14-3   *(Continued)*

NFPA 1500
FIRE DEPARTMENT OCCUPATIONAL SAFETY AND HEALTH PROGRAM WORKSHEET

Standard Content	New in 2002 Ed.	Compliance	Partial Compliance	Compliance with Admin. Action	Expected Compliance Date	Compliance Date	Compliance with Fiscal Action	Estimated $$	Targeted Compliance Date	Remarks or Modifications
4.17 Occupational Safety and Health Committee. Safety and health committee is established										
4.18 Infection Control										
4.18.1 NFPA 1581, fire dept. infection control										
4.18.2 The health and safety officer maintains a liaison with the person or persons designated as infection control officer										
4.18.3 The health and safety officer functions as the fire dept. infection control officer if an infection control officer position does not exist in the fire dept.	✓									
4.19 Critical Incident Stress Management										
4.19.1 Establishment of a critical incident stress management (CISM) program										
4.19.2 CISM program is incorporated into the member assistance program	✓									
4.20 Post-Incident Analysis										
4.20.1 Develop procedures for post-incident analysis										
4.20.2 Written report										
4.20.3 Includes information from the incident safety officer	✓									
4.20.3.1 Includes the incident action plan and the incident safety officer's incident safety plan	✓									
4.20.4 Protective clothing and equipment, personnel accountability system, rehabilitation operations	✓									
4.20.5 NFPA 1561, emergency services incident management systems	✓									
Chapter 5 Training and Education										
5.1 General Requirements										
5.1.1 Safety and health training										
5.1.2 Training for duties and functions										
5.1.3 Training education programs for new members										

Copyright © 2002 National Fire Protection Association   (Page 6 of 25)

FORM 14-3 *(Continued)*

NFPA 1500
FIRE DEPARTMENT OCCUPATIONAL SAFETY AND HEALTH PROGRAM WORKSHEET

Standard Content	New in 2002 Ed.	Compliance	Partial Compliance	Compliance with Admin. Action	Expected Compliance Date	Compliance Date	Compliance with Fiscal Action	Estimated $$	Targeted Compliance Date	Remarks or Modifications
5.1.4 Restrict the use of new members										
5.1.5 Training on the risk management plan	✓									
5.1.6 Written procedures and guidelines										
5.1.7 Training for emergency medical services										
5.1.8 Operation, limitation, maintenance, and retirement criteria for personal protective equipment	✓									
5.1.9 Proficiency in skills and knowledge										
5.1.10 Training programs for all members engaged in emergency operations										
5.1.11 Incident management and accountability system used by the fire dept.	✓									
5.2 Training Curriculums and Requirements										
5.2.1 NFPA 1001, Fire Fighter I										
5.2.2 NFPA 1002, driver/operator										
5.2.3 NFPA 1003, airport fire fighter										
5.2.4 NFPA 1006, rescue technician										
5.2.5 NFPA 1021, fire officer										
5.2.6 NFPA 1051, wildland fire fighting										
5.2.7 NFPA 472, Hazardous materials responders, all members trained to at least first responder operations level										
5.2.8 NFPA 1581, fire dept. infectious disease control										
5.2.9 Adopt or develop training and education curriculums	✓									
5.2.10 NFPA 1403, live fire training										
5.2.11 Supervised training										
5.2.12 AHJ emergency medical services										
5.2.13 Care, use, inspection, maintenance, and limitations of the protective clothing and equipment	✓									

(Page 7 of 25)

FORM 14-3   *(Continued)*

NFPA 1500
FIRE DEPARTMENT OCCUPATIONAL SAFETY AND HEALTH PROGRAM WORKSHEET

Standard Content	New in 2002 Ed.	Compliance	Partial Compliance	Compliance with Admin. Action	Expected Compliance Date	Compliance with Fiscal Action	Estimated $$	Targeted Compliance Date	Remarks or Modifications
5.3 Training Frequency and Proficiency									
5.3.1 Training for all members as often as necessary									
5.3.2 Proficiency of members									
5.3.3 Monitor training progress									
5.3.4 Annual skills check									
5.3.5 Support minimum qualifications and certifications of members									
5.3.6 Members practice assigned skill sets on a regular basis but not less than annually	✓								
5.3.7 Training to members when written policies, practices, procedures, or guidelines are changed and/or updated	✓								
5.3.8 NFPA 1404, SCBA program	✓								
5.3.9 Wildland fire fighters trained at least annually in the proper deployment of fire shelter									
5.4 Special Operations Training									
5.4.1 Advanced training for special operations									
5.4.2 Train members for support to special operations									
5.4.3 NFPA 472, hazardous materials incidents									
5.4.4 NFPA 1670, operations and training for technical rescue incidents and NFPA 1006, rescue technician	✓								
Chapter 6 Fire Apparatus, Equipment, and Driver/Operators									
6.1 Fire Department Apparatus									
6.1.1 Safety and health concerns									
6.1.1.1 Restraint devices	✓								
6.1.2 NFPA 1901, automotive fire apparatus									
6.1.3 NFPA 1906, wildland fire apparatus									
6.1.4 NFPA 1925, marine fire-fighting vessels	✓								

(Page 8 of 25)

FORM 14-3   *(Continued)*

NFPA 1500
FIRE DEPARTMENT OCCUPATIONAL SAFETY AND HEALTH PROGRAM WORKSHEET

Standard Content	New in 2002 Ed.	Compliance	Partial Compliance	Compliance with Admin. Action	Expected Compliance Date	Compliance Date	Compliance with Fiscal Action	Estimated $$	Targeted Compliance Date	Remarks or Modifications
6.1.5 Secure tools, equipment, and SCBA										
6.1.6 NFPA 1912, apparatus refurbishing	✓									
6.1.7 Aircraft provides 4-point restraints for pilots and passengers	✓									
6.1.8 Members performing hoist rescue in aircraft secured by a safety harness	✓									
6.2 Drivers/Operators of Fire Department Apparatus										
6.2.1 Successful completion of approved driver training										
6.2.2 Valid driver's licenses										
6.2.2.1 Traffic laws										
6.2.3 Rules and regulations for private vehicles for emergency response	✓									
6.2.3.1 Rules and regulations for fire department vehicles										
6.2.4 Drivers are responsible										
6.2.4.1 Officers also assume responsibility										
6.2.5 All persons secured										
6.2.6 Drivers obey all traffic laws										
6.2.7 SOPs for non-emergency and emergency response										
6.2.7.1 Safe arrival										
6.2.8 Emergency response, drivers bring vehicle to a complete stop										
6.2.9 Proceed only when safe										
6.2.10 Stop at unguarded railroad grade crossings										
6.2.11 Use caution at guarded railroad grade crossing										
6.2.12 SOPs – engine, transmission and driveline retarders										
6.2.13 SOPs – manual brake limiting valves										
6.3 Riding in Fire Apparatus										
6.3.1 Tail steps and standing prohibited										
6.3.2 Seat belts shall not be released while the vehicle is in motion										

(Page 9 of 25)

FORM 14-3 *(Continued)*

NFPA 1500
FIRE DEPARTMENT OCCUPATIONAL SAFETY AND HEALTH PROGRAM WORKSHEET

Standard Content	New in 2002 Ed.	Compliance	Partial Compliance	Compliance with Admin. Action	Expected Compliance Date	Compliance Date with Fiscal Action	Estimated $$	Targeted Compliance Date	Remarks or Modifications
6.1.5 Secure tools, equipment, and SCBA									
6.1.6 NFPA 1912, apparatus refurbishing	✓								
6.1.7 Aircraft provides 4-point restraints for pilots and passengers	✓								
6.1.8 Members performing hoist rescue in aircraft secured by a safety harness	✓								
6.2 Drivers/Operators of Fire Department Apparatus									
6.2.1 Successful completion of approved driver training									
6.2.2 Valid driver's licenses									
6.2.2.1 Traffic laws									
6.2.3 Rules and regulations for private vehicles for emergency response	✓								
6.2.3.1 Rules and regulations for fire department vehicles									
6.2.4 Drivers are responsible									
6.2.4.1 Officers also assume responsibility									
6.2.5 All persons secured									
6.2.6 Drivers obey all traffic laws									
6.2.7 SOPs for non-emergency and emergency response									
6.2.7.1 Safe arrival									
6.2.8 Emergency response, drivers bring vehicle to a complete stop									
6.2.9 Proceed only when safe									
6.2.10 Stop at unguarded railroad grade crossings									
6.2.11 Use caution at guarded railroad grade crossing									
6.2.12 SOPs – engine, transmission and driveline retarders									
6.2.13 SOPs – manual brake limiting valves									
6.3 Riding in Fire Apparatus									
6.3.1 Tail steps and standing prohibited									
6.3.2 Seat belts shall not be released while the vehicle is in motion									

Copyright © 2002 National Fire Protection Association   (Page 9 of 25)

FORM 14-3   *(Continued)*

NFPA 1500
FIRE DEPARTMENT OCCUPATIONAL SAFETY AND HEALTH PROGRAM WORKSHEET

Standard Content	New in 2002 Ed.	Compliance	Partial Compliance	Compliance with Admin. Action	Expected Compliance Date	Compliance with Fiscal Action	Estimated $$	Targeted Compliance Date	Remarks or Modifications
6.5.8 Tested at least annually									
6.5.9 Remove from service									
6.5.10 NFPA 1581, fire dept. infection control									
6.5.11 NFPA 1932, fire dept. ground ladders									
6.5.12 NFPA 1962, fire hose									
6.5.13 NFPA 10, portable fire extinguishers									
6.5.14 NFPA 1936, powered rescue tools									
Chapter 7 Protective Clothing and Protective Equipment									
7.1 General									
7.1.1 Fire dept. provides PPE									
7 1.2 Use of PPE									
7.1.3 NFPA 1851, selection care and maintenance	✓								
7.1.4 Proper cleaning	✓								
7.1.5 NFPA 1975, work uniforms									
7.1.6 Avoid wearing any clothing that is considered unsafe									
7.1.7 Laundry service for contaminated clothing									
7.1.7.1 Proper cleaning procedures									
7.1.7.2 Washing machines for protective or work clothing									
7.2 Protective Clothing for Structural Fire Fighting									
7.2.1 NFPA 1971, protective clothing									
7.2.2 Minimum 5.08-cm (2-in.) overlap of all protective clothing layers	✓								
7.2.2.1 Garments measured without SCBA	✓								
7.2.3 Overlap not required on continuous composite protection coveralls	✓								
7.2.4 Protective resilient wristlets provided									

(Page 11 of 25)

FORM 14-3   *(Continued)*

NFPA 1500
FIRE DEPARTMENT OCCUPATIONAL SAFETY AND HEALTH PROGRAM WORKSHEET

Standard Content	New in 2002 Ed.	Compliance	Partial Compliance	Compliance with Admin. Action	Expected Compliance Date	Compliance Date Fiscal Action	Compliance with Fiscal Action	Estimated $$	Targeted Compliance Date	Remarks or Modifications
7.2.5 Maintenance of clothing and equipment										
7.2.5.1 Establish a maintenance and inspection program	✓									
7.2.5.2 Assign responsibilities for inspection and maintenance	✓									
7.2.6 Require all members to wear all the protective ensemble										
7.3 Protective Clothing for Proximity Fire-Fighting Operations										
7.3.1 Proximity fire fighting shall be provided with and shall use proximity protective equipment										
7.3.1.1 NFPA 1976, proximity protective clothing	✓									
7.3.2 Minimum 5.08-cm (2-in.) overlap of all proximity protective clothing layers										
7.3.2.1 Garments measured without SCBA	✓									
7.3.3 Overlap not required on continuous full thermal and radiant heat protective coveralls										
7.3.4 Failure of the SCBA										
7.3.4.1 Radiant reflective criteria over SCBA, worn over the outside of proximity protective clothing										
7.4 Protective Clothing for Emergency Medical Operations										
7.4.1 NFPA 1999, emergency medical protective clothing										
7.4.2 Members shall not initiate patient care before the emergency medical gloves are in place	✓									
7.4.3 Fire fighters likely to be exposed to airborne infectious disease provided with NIOSH-approved Type C respirators	✓									
7.4.4 Members shall use emergency medical body and face protection										
7.4.5 NFPA 1581, infection control program protective clothing cleaning										

(Page 12 of 25)

FORM 14-3   *(Continued)*

NFPA 1500
FIRE DEPARTMENT OCCUPATIONAL SAFETY AND HEALTH PROGRAM WORKSHEET

Standard Content	New in 2002 Ed.	Compliance	Partial Compliance	Compliance with Admin. Action	Expected Compliance Date	Compliance Date	Compliance with Fiscal Action	Estimated $$	Targeted Compliance Date	Remarks or Modifications
7.5 Chemical-Protective Clothing for Hazardous Material Emergency Operations										
7.5.1 Vapor-Protective Garments										
7.5.1.1 Members provided with and shall use vapor-protective suits										
7.5.1.2 NFPA 1991, vapor-protective suits	✓									
7.5.1.3 Garment appropriate for the specific hazardous materials emergency										
7.5.1.4 SCBA during hazardous chemical emergency										
7.5.1.4.1 Additional outside NIOSH-approved air supplies permitted	✓									
7.5.1.5 Use only in vapor hazard atmospheres	✓									
7.5.1.6 Use for protection from liquid splash or solid chemicals and particulate protection permitted										
7.5.2 Liquid Splash-Protective Garments										
7.5.2.1 Members provided with and shall use liquid splash-protective suits										
7.5.2.2 NFPA 1992, liquid splash-protective suits	✓									
7.5.2.3 Garment is appropriate for the specific hazardous chemical emergency										
7.5.2.4 Respiratory protection	✓									
7.5.2.4.1 Additional outside NIOSH-approved air supplies permitted	✓									
7.5.2.5 Use for protection from chemicals in vapor form or from unknown liquid chemicals or chemical mixtures prohibited										
7.5.2.5.1 Only vapor-protective suits specified shall be considered for use	✓									
7.5.2.6 Not for carcinogen										
7.5.2.7 Not for toxins										

(Page 13 of 25)

FORM 14-3 *(Continued)*

NFPA 1500
FIRE DEPARTMENT OCCUPATIONAL SAFETY AND HEALTH PROGRAM WORKSHEET

Standard Content	New in 2002 Ed.	Compliance	Partial Compliance	Compliance with Admin. Action	Expected Compliance Date	Compliance Date	Compliance with Fiscal Action	Estimated $$	Targeted Compliance Date	Remarks or Modifications
7.5.2.8 Use only for liquid splash protection										
7.5.2.9 Use for protection from solid chemicals and particulate permitted										
7.6 Inspection, Maintenance, and Disposal of Chemical-Protective Clothing										
7.6.1 Manufacturer's recommendations										
7.6.2 Dispose of contaminated garments										
7.6.2.1 State or federal regulations										
7.8 Protective Clothing and Equipment for Wildland Fire Fighting										
7.8.1 Operating guidelines										
7.8.2 NFPA 1977, protective clothing										
7.8.3 NFPA 1977, protective helmet										
7.8.4 NFPA 1977, protective gloves										
7.8.5 NFPA 1977, protective footwear										
7.8.6 Members provided with an approved fire shelter										
7.9 Respiratory Protection Program										
7.9.1 Respiratory protection program that addresses the selection, care, maintenance, and use	✓									
7.9.2 SOPs that are compliant with this standard	✓									
7.9.3 Members certified for equipment use at least annually	✓									
7.9.4 Reserve SCBA	✓									
7.9.5 Adequate reserve air supply	✓									
7.9.6 Ready-for-use	✓									
7.9.7 Provide and use NFPA 1981, open-circuit self-contained breathing apparatus	✓									
(1) Hazardous atmospheres	✓									
(2) Suspected hazardous	✓									
(3) Can become hazardous	✓									
7.9.8 Keep facepiece in place	✓									

     (Page 14 of 25)

FORM 14-3   (*Continued*)

NFPA 1500
FIRE DEPARTMENT OCCUPATIONAL SAFETY AND HEALTH PROGRAM WORKSHEET

Standard Content	New in 2002 Ed.	Compliance	Partial Compliance	Compliance with Admin. Action	Expected Compliance Date	Compliance with Fiscal Action	Estimated $$	Targeted Compliance Date	Remarks or Modifications
7.10 Breathing Air									
7.10.1 Grade D air: ANSI/CGA G7.1	✓								
7.10.2 Certification and documentation for vendor-provided air	✓								
7.10.3 Fire dept. manufactures own—tested every 3 months	✓								
7.10.4 Fire dept. obtains own—tested every 3 months	✓								
7.10.5 Fire dept. at point of transfer from the storage cylinders—tested every 3 months	✓								
7.11 Respiratory Protection Equipment									
7.11.1 SCBA	✓								
7.11.1.1 NFPA 1981, open-circuit self-contained breathing apparatus, 1987 edition or later	✓								
7. 11 1.2 Closed-circuit SCBA permitted	✓								
7.11.1.3 Closed-circuit SCBA NIOSH-certified rated service life of at least 2 hours and positive-pressure only	✓								
7.11.2 Supplied-Air Respirators	✓								
7.11.2.1 AHJ air respirator	✓								
7.11.2.2 Not used in IDLH atmospheres unless equipped with a NIOSH-certified escape air cylinder and a pressure-demand facepiece	✓								
7.11.2.3 Type C Pressure-Demand Class not used in IDLH atmospheres unless they meet manufacturer's specifications	✓								
7.11.3 Full Facepiece Air-Purifying Respirators	✓								
7.11.3.1 Only used in non-IDLH atmospheres for contaminants NIOSH certifies	✓								
7.11.3.2 AHJ provide NIOSH-certified respirators	✓								
7.11.3.3 AHJ establish a policy for service life	✓								

Copyright © 2002 National Fire Protection Association   (Page 15 of 25)

FORM 14-3   (Continued)

NFPA 1500
FIRE DEPARTMENT OCCUPATIONAL SAFETY AND HEALTH PROGRAM WORKSHEET

Standard Content	New in 2002 Ed.	Compliance	Partial Compliance	Compliance with Admin. Action	Expected Compliance Date	Compliance with Fiscal Action	Estimated $$	Targeted Compliance Date	Remarks or Modifications
7.12 Fit Testing									
7.12.1 Facepiece seal qualitative or quantitative fit test annually	✓								
7.12.2 Only members with a properly fitting facepiece permitted function in a hazardous atmosphere	✓								
7.12.3 Respirators quantitative or qualitative fit testing	✓								
7.12.4 AHJ-required test protocols	✓								
7.12.5 Records of facepiece fitting tests	✓								
7.12.6 Protection factor at least 500 for negative-pressure facepieces	✓								
7.12.7 Facepiece face seal	✓								
7.12.8 Beard and facial hair	✓								
7.12.8.1 Restrictions apply regardless of test	✓								
7.12.9 Spectacles	✓								
7.12.10 Spectacle strap or temple bars prohibited									
7.12.11 Contact lens permitted									
7.12.12 Head covering breaking seal prohibited									
7.12.13 SCBA facepiece/head harness worn under protective hood	✓								
7.12.14 SCBA facepiece/head harness worn under hazardous chemical protective clothing helmet	✓								
7.12.15 Helmets shall not interfere with the respiratory protection facepiece-to-face seal									
7.13 SCBA Cylinders									
7.13.2 Hydrostatic test cylinders	✓								
7.13.3 In-service SCBA cylinders	✓								
7.13.4 In-service SCBA cylinders inspected weekly, monthly, and prior to filling	✓								
7.13.5 Filling SCBA cylinders personnel shall be protected	✓								
7.13.6 Rapid filling	✓								
7.13.7 Risk assessment	✓								

(Page 16 of 25)

FORM 14-3   (Continued)

NFPA 1500
FIRE DEPARTMENT OCCUPATIONAL SAFETY AND HEALTH PROGRAM WORKSHEET

Standard Content	New in 2002 Ed.	Compliance	Partial Compliance	Compliance with Admin. Action	Expected Compliance Date	Compliance with Fiscal Action	Estimated $$	Targeted Compliance Date	Remarks or Modifications
7.13.8 Rapid refilling SCBA	✓								
7.13.9 Emergency situation air transfer permitted	✓								
7.13.10 Transfilling manufacturer's instructions	✓								
7.14 Personal Alert Safety System (PASS)									
7.14.1 NFPA 1982, personal alert safety systems (PASS)									
7.14.2 Members provided with and use PASS device	✓								
7.14.3 Tested at least weekly and prior to use									
7.15 Life Safety Rope and System Components									
7.15.1 NFPA 1983, life safety rope and system components									
7.15.2 Life safety rope									
7.15.2.1 Remove from service									
7.15.3 Life safety rope inspection before reuse:									
(1) No damage from fires, chemicals, or abrasives									
(2) No impact load									
(3) No materials known to deteriorate ropes									
7.15.3.1 Destroyed after emergency use if failed 7.15.3									
7.15.3.2 Removed from service									
7.15.4 Other use removed from service and destroyed									
7.15.5 Inspection by qualified inspector									
7.15.6 Records for each life safety rope used at incident/training									
7.16 Face and Eye Protection									
7.16.1 Eye protection appropriate for hazard									
7.16.1.1 ANSI Z87.1, practice for occupational and educational eye and face protection									

(Page 17 of 25)

FORM 14-3 *(Continued)*

NFPA 1500
FIRE DEPARTMENT OCCUPATIONAL SAFETY AND HEALTH PROGRAM WORKSHEET

Standard Content	New in 2002 Ed.	Compliance	Partial Compliance	Compliance with Admin. Action	Expected Compliance Date	Compliance Date Fiscal Action	Compliance with Estimated $$	Targeted Compliance Date	Remarks or Modifications
7.16.2 SCBA facepieces—primary face and eye protection									
7.16.2.1 Facepiece-mounted regulator	✓								
7.16.3 Helmet face shield partial face protection	✓								
7.17 Hearing Protection									
7.17.1 Use in excess of 90 dBA apparatus									
7.17.2 Use in excess of 90 dBA tools and equipment									
7.17.3 Hearing conservation program									
7.18 New and Existing Protective Clothing and Protective Equipment									
7.18.1 New PPE meets current standards									
7.18.2 Existing PPE shall have met standards when manufactured									
7.18.3 PPE manufactured prior to the 1986 edition of NFPA 1971 shall be removed from service	✓								
Chapter 8 Emergency Operations									
8.1 Incident Management									
8.1.1 Prevent accidents and injuries									
8.1.2 NFPA 1561, incident management system in writing	✓								
8.1.3 Used at all emergency incidents									
8.1.4 IMS applied to drills, exercises for training									
8.1.5 IC incident responsible for safety									
8.1.6 Incident safety officer									
8.1.7 Span of control									
8.1.8 IC incident responsibility									
8.1.9 NFPA 1561, fire dispatch and fireground communication and NFPA 1221, emergency communication systems	✓								
8.1.10 SOPs use of clear text radio messages	✓								
8.1.10.1 SOPs use "emergency traffic" to clear radio traffic	✓								
8.1.10.2 "Emergency traffic" permitted	✓								

Copyright © 2002 National Fire Protection Association   (Page 18 of 25)

FORM 14-3   *(Continued)*

NFPA 1500
FIRE DEPARTMENT OCCUPATIONAL SAFETY AND HEALTH PROGRAM WORKSHEET

Standard Content	New in 2002 Ed.	Compliance	Partial Compliance	Compliance with Admin. Action	Expected Compliance Date	Compliance Date	Compliance with Fiscal Action	Estimated $$	Targeted Compliance Date	Remarks or Modifications
8.1.11 Identify emergency	✓									
8.1.11.1 All clear	✓									
8.1.12 Incident clock	✓									
8.1.12.1 Dispatch notify IC	✓									
8.1.12.2 IC cancel the incident clock	✓									
8.2 Risk Management During Emergency Operations										
8.2.1 Risk management in incident command										
8.2.2 Risk management principles										
8.2.3 Elevating members' risks										
8.2.3.1 Limit to defensive operations	✓									
8.2.4 Risk management principles define limits of acceptable/ unacceptable position/functions										
8.2.5 Qualified personnel — safety of operations										
8.2.6 IC to ensure body armor available for civil disturbances	✓									
8.2.7 Incidents involving risk of physical violence	✓									
8.2.8 Nerve agents	✓									
8.3 Personnel Accountability During Emergency Operations										
8.3.1 Written SOP—NFPA 1561, incident management system										
8.3.2 Local conditions and characteristics										
8.3.3 Members actively participate										
8.3.4 IC maintain awareness										
8.3.5 Sector officers responsible	✓									
8.3.6 Company officers responsible	✓									
8.3.7 Fire fighters remain with company	✓									
8.3.8 Fire fighters responsible	✓									
8.3.9 Used all incidents	✓									
8.3.10 Accountability system effective	✓									
8.3.11 Additional accountability officers	✓									
8.3.12 Tracking and accountability of assigned companies	✓									

(Page 19 of 25)

FORM 14-3   *(Continued)*

NFPA 1500
FIRE DEPARTMENT OCCUPATIONAL SAFETY AND HEALTH PROGRAM WORKSHEET

Standard Content	New in 2002 Ed.	Compliance	Partial Compliance	Compliance with Admin. Action	Expected Compliance Date	Compliance Date	Fiscal Action with	Estimated $$	Targeted Compliance Date	Remarks or Modifications
8.4 Members Operating at Emergency Incidents										
8.4.1 The fire department shall provide an adequate number of personnel to safely conduct emergency scene operations										
8.4.1.1 Safe operations										
8.4.2 Established safety criteria										
8.4.3 Direct supervision										
8.4.3.1 The requirement shall not reduce the training requirements	✓									
8.4.4 Teams of two or more										
8.4.5 Crew members operating shall be in communication										
8.4.6 Crew members shall be in proximity to each other										
8.4.7 Initial state—one team assigned-standby person										
8.4.8 Standby members responsible										
8.4.9 Standby members remain in communication										
8.4.10 Initial stage										
8.4.11 Standby member permitted to perform other duties outside of the hazard area										
8.4.12 Standby member permitted activities										
8.4.12.1 Standby member not permitted										
8.4.13 Full protective clothing, protective equipment, and SCBA										
8.4.13.1 Full protective clothing, protective equipment accessible										
8.4.14 Standby members shall don full protective clothing, protective equipment, and SCBA										
8.4.15 Single team assignment, one rapid intervention crew										
8.4.16 Second team assignment, one rapid intervention crew										

FORM 14-3   (Continued)

NFPA 1500
FIRE DEPARTMENT OCCUPATIONAL SAFETY AND HEALTH PROGRAM WORKSHEET

Standard Content	New in 2002 Ed.	Compliance	Partial Compliance	Compliance with Admin. Action	Expected Compliance Date	Compliance Date Fiscal Action	Compliance with Fiscal Action	Estimated $$	Targeted Compliance Date	Remarks or Modifications
8.4.17 In imminent life-threatening situation, action to prevent loss of life permitted with less than four personnel										
8.4.17.1 No exception shall be permitted when there is no possibility to save lives										
8.4.17.2 Actions taken investigated										
8.4.18 Aircraft rescue fire fighting, IDLH area within 23 m (75 ft) of aircraft	✓									
8.4.19 IDLH adjusts to meet operational needs	✓									
8.4.20 Inside IDLH area	✓									
8.4.21 Highest available level of emergency medical care for special operations—basic life support minimum										
8.4.22 NFPA 473, EMS for hazardous materials operations										
8.4.23 Basic life support for other emergency operations										
8.4.24 Secured to aerial device										
8.4.25 Fluorescent retroreflective material—MV traffic										
8.4.26 Apparatus utilized as a shield										
8.4.27 Warning device for oncoming traffic	✓									
8.4.28 SCBA for arson investigators in IDLH atmosphere	✓									
8.4.29 Water rescue members wear flotation devices	✓									
8.5 Rapid Intervention for Rescue of Members										
8.5.1 Rescue of members										
8.5.2 Rapid intervention crew										
8.5.2.1 Fully equipped—PPE, SCBA, rescue equipment										
8.5.3 Composure and structure										
8.5.4 IC evaluate risks to crews										
8.5.5 Crew(s) status—early stages	✓									
8.5.6 Crew(s) status—expanded incident	✓									
8.5.7 Special operations rapid intervention crew(s)										

Copyright © 2002 National Fire Protection Association   (Page 21 of 25)

FORM 14-3   *(Continued)*

NFPA 1500
FIRE DEPARTMENT OCCUPATIONAL SAFETY AND HEALTH PROGRAM WORKSHEET

Standard Content	New in 2002 Ed.	Compliance	Partial Compliance	Compliance with Admin. Action	Expected Compliance Date	Compliance Date	Compliance with Fiscal Action	Estimated $$	Targeted Compliance Date	Remarks or Modifications
8.6 Rehabilitation During Emergency Operations										
8.6.1 SOP for rehabilitation of members										
8.6.2 IC provide rest and rehab										
8.6.3 On-scene rehabilitation to include basic life support										
8.6.4 Each member responsible to communicate rest and rehab needs	✓									
8.6.5 Each wildland fire fighter provided with 2 L (2 qt) of water										
8.6.5.1 Replenishment of water supplies										
8.7 Civil Unrest/Terrorism										
8.7.1 Fire dept. not involved in crowd control	✓									
8.7.2 SOPs—civil disturbance	✓									
8.7.2.1 Situations of disturbances	✓									
8.7.3 Interagency agreement	✓									
8.7.4 Indication of life and death situation requiring law enforcement intervention	✓									
8.7.5 Coordinate with law enforcement IC	✓									
8.7.6 Fire dept. IC identify and react to violent situations	✓									
8.7.7 Fire dept. IC communicate with law enforcement IC	✓									
8.7.8 Stage resources in a safe area	✓									
8.7.9 Secure law enforcement when violence occurs	✓									
8.7.10 Fire dept. support to SWAT	✓									
8.7.11 Special SOPs for members	✓									
8.7.11.1 Special operations										
8.8 Post-Incident Analysis										
8.8.1 SOPs—post-incident critique										
8.8.2 Critique involves incident safety officer	✓									
8.8.3 Basic review on the safety and health of members										
8.8.4 Identify needed action	✓									
8.8.5 Standard action plan	✓									
8.8.5.1 Includes change needed and the responsibilities	✓									

Copyright © 2002 National Fire Protection Association (Page 22 of 25)

FORM 14-3 *(Continued)*

NFPA 1500
FIRE DEPARTMENT OCCUPATIONAL SAFETY AND HEALTH PROGRAM WORKSHEET

Standard Content	New in 2002 Ed.	Compliance	Partial Compliance	Compliance with Admin. Action	Expected Compliance Date	Compliance Date with Fiscal Action	Estimated $$	Targeted Compliance Date	Remarks or Modifications
Chapter 9 Facility Safety									
9.1 Safety Standards									
9.1.1 Comply with codes									
9.1.2 NFPA 1581, infection control									
9.1.3 All facilities—smoke detectors									
9.1.3.1 Interconnected detectors									
9.1.4 All facilities—carbon monoxide detectors									
9.1.5 New/existing facilities comply with *Life Safety Code*									
9.1.6 Prevent exhaust exposure	✓								
9.1.7 Smokefree facilities									
9.2 Inspections									
9.2.1 Annual code inspection									
9.2.2 Inspections documented and recorded									
9.2.3 Monthly safety inspection									
9.3 Maintenance and Repairs. Established system									
Chapter 10 Medical and Physical Requirements									
10.1 Medical Requirements									
10.1.1 Medical evaluation and certification before becoming a member									
10.1.2 Risks and functions associated duties	✓								
10.1.3 NFPA 1582, medical requirements	✓								
10.1.4 Aircraft pilots comply with FAA regulations	✓								
10.1.5 Under the influence of alcohol or drugs	✓								
10.2 Physical Performance Requirements									
10.2.1 Fire dept. develop requirements									
10.2.2 Certification for use of respiratory protection conducted annually									
10.2.3 Candidated certified by fire dept.									
10.2.4 Current fire fighters annually certified by fire dept.									

(Page 23 of 25)

FORM 14-3 (*Continued*)

NFPA 1500
FIRE DEPARTMENT OCCUPATIONAL SAFETY AND HEALTH PROGRAM WORKSHEET

Standard Content	New in 2002 Ed.	Compliance	Partial Compliance	Compliance with Admin. Action	Expected Compliance Date	Compliance Date with Fiscal Action	Estimated $$	Targeted Compliance Date	Remarks or Modifications
10.2.5　Members not be permitted in emergency operations									
10.2.6　Physical performance rehabilitation	✓								
10.3　Health and Fitness									
10.3.1　NFPA 1583, fitness programs	✓								
10.3.2　Fitness levels determined individual's assigned functions									
10.3.3　Health and fitness coordinator administer the program									
10.3.4　Health and fitness coordinator act as liaison									
10.4　Confidential Health Data Base									
10.4.1　Confidential permanent health file									
10.4.2　File records the results evaluations									
10.4.3　Individual/group records									
10.4.4　Record autopsy results	✓								
10.5　Infection Control									
10.5.1　Fire dept. limit or prevent member's exposure									
10.5.2　NFPA 1581, infection control									
10.6　Fire Department Physician									
10.6.1　Fire dept. physician									
10.6.2　Provide medical guidance									
10.6.3　Licensed doctor									
10.6.4　Availablity									
10.6.4.1　Availability permitted access to a number physicians	✓								
10.6.5　Health and safety officer and health fitness coordinator liaison with physician	✓								
Chapter 11　Member Assistance and Wellness Programs									
11.1　Member assistance program									
11.1.1　Provide member assistance program									
11.1.2　Member and family assistance program									

Copyright © 2002 National Fire Protection Association　　　　　　　　　(Page 24 of 25)

FORM 14-3　(Continued)

NFPA 1500
FIRE DEPARTMENT OCCUPATIONAL SAFETY AND HEALTH PROGRAM WORKSHEET

Standard Content	New in 2002 Ed.	Compliance	Partial Compliance	Compliance with Admin. Action	Expected Compliance Date	Compliance with Fiscal Action	Estimated $$	Targeted Compliance Date	Remarks or Modifications
11.1.3 Written policy									
11.1.4 Written rules for records									
11.1.4.1 Rules for conditions of records released	✓								
11.1.5 Records maintained not part of member's personnel file	✓								
11.2 Wellness Program									
11.2.1 Wellness program									
11.2.2 Health promotion activities									
11.2.2.1 Provide a smoking/tobacco use cessation program	✓								
Chapter 12 Critical Incident Stress Program									
12.1 General									
12.1.1 Physician to provide guidance									
12.1.2 Written policy—program to relieve stress									
12.1.3 Criteria for implementation									
12.1.4 Program available to members for situations involving psychological and physical well-being	✓								

(Page 25 of 25)

FORM 14-3   *(Continued)*

Fire Department Infectious Exposure Form

Exposed member's name: _____   Rank: _____

Soc. Sec. No.: _____

Field IncNo.:. _____   Shift: _____   Company: _____   District: _____

Name of patient: _____   Sex: _____

Age: _____   Address: _____

Suspected or confirmed disease: _____

Transported to: _____

Transported by: _____

Date of exposure: _____   Time of exposure: _____

Type of incident (auto accident, trauma): _____

What were you exposed to?

❏ Blood   ❏ Tears   ❏ Feces   ❏ Urine   ❏ Saliva   ❏ Vomitus   ❏ Sputum   ❏ Sweat

❏ Other _____

What part(s) of your body became exposed? Be specific: _____

_____

_____

Did you have any open cuts, sores, or rashes that became exposed? Be specific: _____

_____

_____

How did exposure occur? Be specific: _____

_____

_____

Did you seek medical attention?   ❏ yes   ❏ no

Where? _____   Date: _____

Contacted infection control officer? Date: _____   Time: _____

Supervisor's signature: _____   Date: _____

Member's signature: _____   Date: _____

(Page 1 of 1)

FORM 14-4   Sample Fire Department Infectious Exposure Form

Source: NFPA 1581, *Standard on Fire Department Infection Control Program*, 2000 ed., Figure A-2-6.4

15

Line-of-Duty Death Resources for Fire Departments

To help a fire department deal with the death of a fire fighter in the line of duty, the National Fallen Firefighters Foundation provides an array of resources and guidelines. This chapter offers some of this information, including the following:

- First steps to help the family and the department
- Information on resources, training, and standard operating procedures
- Line-of-duty death action checklist

For additional i nformation, contact the National Fallen Firefighters Foundation at 301-447-1365 or go to their website: http://www.firehero.com.

FIRST STEPS

If your fire department has lost a fire fighter in the line of duty, here are a few steps the National Fallen Firefighters Foundation has outlined for your department to take to help both the family and the department.

1. Immediately contact the Department of Justice's Public Safety Officers' Benefits (PSOB) Program at (888) 744-6513. When you report a fire fighter death, have all critical information available on the fire fighter, the department, and the next of kin. PSOB offers financial assistance to survivors of public safety officers who die in the line of duty from a traumatic injury. There are many procedures that need to be followed so survivors can receive benefits to which they are entitled. Call PSOB even if you are not sure whether your fire fighter's family will qualify for benefits under this program. This initial phone call will begin the process of de-termining the survivors' eligibility for benefits.

2. Consult applicable resources. Based on suggestions from chiefs who have lost fire fighters in the line of duty, the National Fallen Firefighters Foundation has created

Source: Information in this chapter reproduced by permission of the National Fallen Firefighters Foundation from their website http://www.firehero.org.

a checklist of what needs to be done immediately, before the funeral, and afterwards. Other line-of-duty death information, including autopsy guidelines, is accessible through the foundation's website. The foundation can also provide suggestions about how to support the family and coworkers during this difficult time. If you would like to speak directly with another chief who has also lost a fire fighter in the line of duty, contact the National Fallen Firefighters Foundation at 301-447-1365.

3. Find out what benefits exist for survivors of fallen fire fighters in your state. Benefits may include lump-sum death payments, workers' compensation, funeral benefits, pensions and retirement programs, scholarships, and nonprofit/private support.

4. Request a copy of the comprehensive *Resource Guide* of line-of-duty death materials for fire departments, which is available from the foundation free of charge. This 50-page guide is intended for pre-incident planning, but it contains information on family and fire department support and resources that may be helpful to you at this time.

RESOURCES

The National Fallen Firefighters Foundation has compiled the following list of resources based upon recommendations from fire departments and grief specialists. The foundation continues to add resources to this list.

Resource Guide

A comprehensive *Resource Guide* of line-of-duty death materials for fire departments is available from the Foundation. The 50-page guide contains information on pre-incident planning, notification, family and fire department support, and resources for departments.

Autopsy Guide

- *Firefighter Autopsy Protocol*, United States Fire Administration, 1991
  Available online at
  http://www.usfa.fema.gov/downloads/pdf/publications/fa-156.pdf
  Contact: United States Fire Administration at www.usfa.fema.gov
  USFA Publications Center
  16825 S. Seton Avenue
  Emmitsburg, MD 21727
  1-800-561-3356

Benefits

- *Public Safety Officers' Benefits Program Fact Sheet*
  Available online at www.ncjrs.org/pdffiles1/bja/fs000271.pdf
- *Public Safety Officers' Educational Assistance Program*
  Available online at www.ncjrs.org/pdffiles1/bja/fs000270.pdf
- *National Fallen Firefighters Foundation, State Benefits*
  Available online at www.firehero.org. See State Benefits section.

Funeral Guides

- *Chaplain's Manual: Fire Department Funerals.* Federation of Fire Chaplains, 1994
  Contact: Federation of Fire Chaplains at
  www.emergency-world.com/chaplains
  E-mail: chapdir1@aol.com
  Route 1, Box 155B, Clifton, Texas 76634
  (254) 622-8514

- *Final Farewell to a Fallen Firefighter: A Basic Fire Department Funeral Protocol.*
  Fire Engineering Magazine, 1993
  Contact: Fire Engineering Magazine
  Email: williamm@pennwell.com for reprint
  Park 80 West, Plaza Two, 7th Floor, Saddle Brook, NJ 07663
  (201) 845-0800/ FAX: (201) 845-6275

- *For Those Who Gave So Much: Planning, Preparation, and Officiation of Funerals
  and Memorial Services for Public Safety Officers.* Dwaine Booth, 1993
  Contact: Booth/Taylor Publishing
  2579 Surrey Drive, Clearwater, FL 34615
  (727) 789-3816

- *Funeral Procedures for Firefighters.* National Volunteer Fire Council, 1991
  Contact: NVFC at www.nvfc.org/manuals.html
  E-mail: nvfcoffice@nvfc.org
  1050 17th Street, NW, Suite 490, Washington, DC 20036
  (202) 887-5700/1-888-ASK-NVFC/FAX: (202) 887-5291

- *An Honorable Farewell.* Warren L. James, Fire Chief Magazine, October 1998
  Contact: Fire Chief Magazine
  35 E. Wacker Drive, Suite 700, Chicago, IL 60601-2198
  (312) 726-7277/FAX: (312) 726-0241

- *IAFF Recommended Protocol for Line-of-Duty Deaths.*
  The IAFF will provide this protocol at the request of the IAFF District Vice President or local IAFF affiliates.
  Contact: IAFF at www.iaff.org
  1750 New York Avenue, NW, Washington, DC 20006
  (202) 737-8484 / FAX: (202) 737-8418

- *Illinois Fire Chiefs Association, Funeral Service Guidelines, Funeral Committee*
  Contact: 800-662-0732, www.illinois.firechiefs.org
  P.O. Box 7, Skokie, IL 60076

- *A Labor of Love, How to Write a Eulogy.* Garry Schaeffer.
  This book, recommended by a fire chief, offers tips on what to say and provides samples.
  Contact: Can be ordered and downloaded online at www.funerals-online.com/1eulogy.htm or call (800) 479-7487. Also available through the National Fallen Firefighters Foundation Lending Library.

- *A Procedural Guide in the Event of Death in the Line of Duty of a Member of the
  Volunteer Fire Service.* National Volunteer Fire Council, 1987.
  Contact: NVFC at www.nvfc.org/manuals.html
  Email: nvfcoffice@nvfc.org
  1050 17th Street, NW, Suite 490, Washington, DC 20036
  (202) 887-5700 / 1-888-ASK-NVFC / FAX: (202) 887-5291

Investigations

- *Guide for Investigation of a Line of Duty Death.* International Association of Fire Chiefs

  Contact: IAFC at www.ichiefs.org
  E-mail: publications@iafc.org
  4025 Fair Ridge Drive, Fairfax, VA 22033-2868
  (703) 273-0911
  Available on line at www.iafc.org/downloads/06Investigations.PDF

- NIOSH Fire Fighter Fatality Programs and Reports

  Firefighter Fatality Investigation Program

  Contact: NIOSH at www.cdc.gov/niosh/implweb.html
  1-800-35-NIOSH or 1-800-356-4674

  Firefighter Fatality Reports

  A list of and links to all the periodic NIOSH reports on firefighter fatalities are available at www.cdc.gov/niosh/facerpts.html

- *The Aftermath of Firefighter Fatality Incidents: Preparing for the Worst.* United States Fire Administration, Technical Report Series, Report 089.

  Contact: United States Fire Administration at www.usfa.fema.gov
  USFA Publications Center, 16825 S. Seton Avenue, Emmitsburg, MD 21727
  1-800-561-3356

Pre-Incident Planning

- *A Guide to Help the Fire Service Prepare for a Line of Duty Death.* Paul J. Antonellis, Jr., 1995

  Contact: Order online at www.fire-police-ems.com/books/bg9500.htm
  Email: fsp@ma.ultranet.com
  FPS Books and Videos, 577 Main Street, Hudson, MA 01749
  1-800-522-8528 or (978) 562-3554

Specialized Websites

Firehouse.com

www.firehouse.com

Provides immediate information on incidents and hosts a forum on line-of-duty death issues.

Provides immediate information on how to report a line-of-duty death and on support for survivors, as well as criteria for inclusion on National Memorial. Gives information on the national tribute held each October. Provides pre-incident planning resources to download and adapt to local needs, as well as a clearinghouse listing honor guard. Lists survivor benefits for each state.

United States Fire Administration

www.usfa.fema.gov/ffmem

Provides listing of firefighter deaths. The current year listings reflect only the information USFA has received and do not indicate that a line-of-duty death will meet criteria for inclusion on the National Memorial.

TRAINING

Thanks to a Department of Justice grant, the National Fallen Firefighters Foundation offers a training program to help senior fire officials prepare for the worst—a line-of-duty death or serious injury. Fire service personnel and families helped develop this one-day course, called "Taking Care of Our Own." It covers pre-incident planning, survivor notification, family and coworker support, and benefits and resources available to the families.

The foundation offers the training at various locations across the country. Check the foundation's website for a schedule of dates and location. The foundation encourages you to look at an overview of the course and to download and use the resources, including the checklists and other information provided in this chapter.

STANDARD OPERATING PROCEDURES FOR LINE-OF-DUTY DEATHS

The National Fallen Firefighters Foundation offers copies of standard operating procedures for line-of-duty deaths developed by fire departments across the country. The SOPs are designed for various sizes and types of department and may be helpful in developing SOPs for your department. These publications are available free of charge.

Tempe Fire Department - Arizona

- Career department
- Department size: 100 to 250
- Size of population served: 100,000 to 200,000

Long Beach Fire Department - California

- Career department
- Department size: 251 to 500
- Size of population served: Over 200,000

Montrose Fire Department - Colorado

- Combination department
- Department size: 26 to 50
- Size of population served: 10,000 to 50,000

Florida Division of Forestry - Florida

- State agency utilizing career and volunteer firefighters

Dixon Fire Department - Illinois

- Combination department
- Department size: 1 to 25
- Size of population served: 10,000 to 50,000

East Alton Fire Department - Illinois

- Combination department
- Department size: 1 to 25
- Size of population served: 10,000 to 50,000
- Includes funeral service guidelines

Charlotte Fire Department - North Carolina

- Career department
- Department size: 500 to 1,000
- Size of population served: Large metropolitan area

Westerville Division of Fire - Ohio

- Career department
- Department size: 100 to 250
- Size of population served: 10,000 to 50,000

Corpus Christi Fire Department - Texas

- Career department
- Department size: 251 to 500
- Size of population served: Over 200,000

North Shore Fire Department - Wisconsin

- Career department
- Department size: 100 to 250
- Size of population served: 65,000

Funeral Procedures

Orange County Fire Department - California

- Combination department
- Department size: Over 1,000
- Size of population served: Large metropolitan county

Chicago Fire Department - Illinois

- Career department
- Department size: Over 1,000
- Size of population served: Large metropolitan area
- Includes administrative duties after a line-of-duty death

Western Missouri Fire Chiefs Association - Missouri

- Funeral policy
- Funeral equipment resource list

LINE-OF-DUTY DEATH CHECKLISTS

The Line-of-Duty Death Action Checklist, which was created by and is available from the National Fallen Firefighters Foundation, is reproduced in Figure 15-1.

LINE-OF-DUTY DEATH ACTION CHECKLIST

FIRST 24 HOURS

Notification

_____   Assign a 2-person team to notify the firefighter's family, in person, before releasing any information.

_____   Notify all on- and off-duty personnel, including chaplain.

_____   Notify elected officials and other key people in the community of the death.

_____   Notify all other fire chiefs in the jurisdiction.

_____   Notify the Public Safety Officers' Benefits Program office.

Family Support

_____   Designate a family support liaison (team) and offer to stay with the family around the clock.

_____   Designate a hospital liaison, if appropriate.

_____   Meet with the family to explain support the fire department can provide and any immediate support they can offer. Be prepared to explain why an autopsy may be required.

_____   Collect the deceased firefighter's department belongings to give to the family later. Inventory and document in the presence of a witness. If some belongings will be held during investigation, explain this to the family.

Department Support

_____   Contact the National Fallen Firefighters Foundation's Chief-to-Chief Network as needed for assistance.

_____   Arrange critical incident debriefing for the department.

Dealing with the Incident

_____   Determine the type of firefighter fatality investigation to conduct in addition to the NIOSH investigation (i.e., internal or external board of inquiry; arson-, accident- or homicide-related).

_____   Contact the departmental or jurisdictional attorney regarding possible legal issues.

Dealing with the Community and the Media

_____   Prepare a summary of facts about the firefighter and the incident to use for public release of information.

_____   Prepare a written statement for the chief or spokesperson to release to the media.

_____   Hold a briefing with the media.

(Page 1 of 2)

FIGURE 15-1   National Fallen Firefighters Foundation Line-of-duty Death Action Checklist

DAY TWO THROUGH THE FUNERAL

Funeral/Memorial Service

_____   Assist the family in planning the funeral as they choose.

_____   Continue to inform department members of the details regarding the incident and the funeral/memorial service plans.

_____   Coordinate plans for fire department participation in funeral.

Family Support

_____   Request that local law enforcement officials make routine checks of the family's residence during the funeral and for several weeks afterwards.

_____   Assist the family with tasks related to home maintenance, transportation of out-of-town family and friends, childcare, etc.

Department Support

_____   Monitor department members closest to the incident to see how they are dealing with the loss.

AFTER THE FUNERAL

Family Support

_____   Continue to invite the family to department events and activities.

_____   Provide assistance with routine tasks (home maintenance, running errands, etc.).

_____   Assign someone to assist the family in accessing all benefits for which they are eligible.

_____   Offer to "be there" at special times/events (children's activities, holidays, etc.).

Department Support

_____   Assist department members in accessing additional support, as needed.

Memorials and Tributes

_____   Inform and include families in local, state, and national tributes to the firefighter.

_____   Make the family aware of the National Fallen Firefighters Foundation and its support programs for fire service survivors.

_____   Plan to attend the National Fallen Firefighters Memorial Weekend and to send an escort and honor guard unit for the family.

Department Issues/Planning

_____   Update Emergency Contact Information for all department members.

_____   Create or revise the department's Line-of-Duty Death plan.

(Page 2 of 2)

FIGURE 15-1   _(Continued)_

16

Sample Incident Management System Forms

This chapter contains full-size reproducible copies of sample incident action planning forms that are routinely used in the incident command system. Proper documentation demands that standard forms be used for all incidents so that all the appropriate information is documented. Forms 16-1 through 16-9 are adapted from the National Inter-Agency Incident Management System's (NIIMS) 1983 publication *Incident Command System,* which is available from Fire Protection Publications, Oklahoma State University at Stillwater. Forms 16-10 through 16-12 were provided by the Phoenix Fire Department. The included forms are

- Form 16-1 Incident Briefing (ICS 201)
- Form 16-2 Incident Objectives (ICS 202)
- Form 16-3 Organization Assignment List (ICS 203)
- Form 16-4 Division Assignment List (ICS 204)
- Form 16-5 Communications Plan (ICS 205)
- Form 16-6 Medical Plan (ICS 206)
- Form 16-7 Incident Organization Chart (ICS 207)
- Form 16-8 Incident Status Summary (ICS 209)
- Form 16-9 Operational Planning Worksheet (ICS 215)
- Form 16-10 EMS Medical Worksheet
- Form 16-11 Tactical Worksheet
- Form 16-12 Post-Incident Review Worksheet

INCIDENT BRIEFING (ICS 201)	1. INCIDENT NAME	2. DATE PREPARED	3. TIME PREPARED

4. MAP SKETCH

ICS 201 Page 1	8. PREPARED BY (NAME AND POSITION)

FORM 16-1　Incident Briefing

Source: NIIMS, ICS 201

7. SUMMARY OF CURRENT ACTIONS

ICS 201
Page 2

FORM 16-1 (*Continued*)

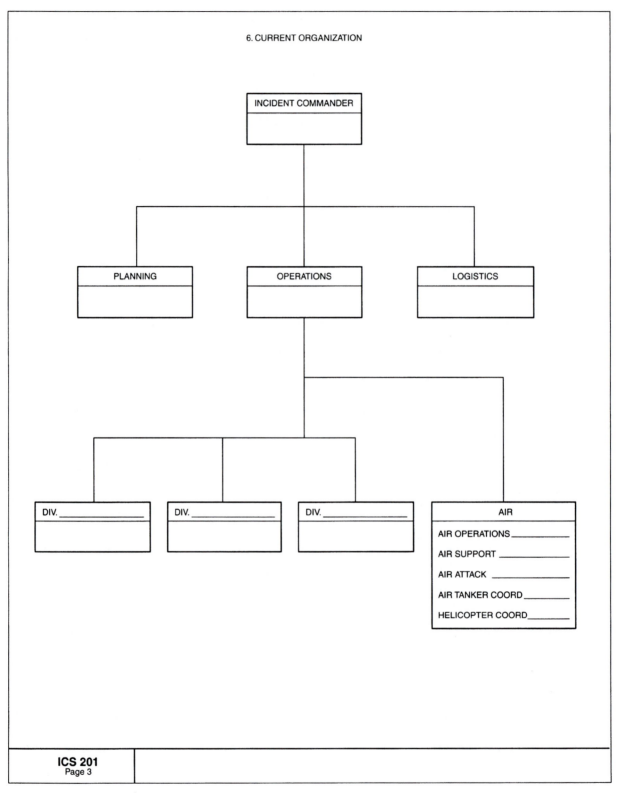

FORM 16-1 *(Continued)*

5. RESOURCES SUMMARY				
RESOURCES ORDERED	RESOURCE IDENTIFICATION	ETA	ON SCENE ✓	LOCATION / ASSIGNMENT
ICS 201 Page 4				

FORM 16-1 (*Continued*)

INCIDENT OBJECTIVES (ICS 202)	1. INCIDENT NAME	2. DATE PREPARED	3. TIME PREPARED

4. OPERATIONAL PERIOD (DATE/TIME)

5. GENERAL CONTROL OBJECTIVES FOR THE INCIDENT (INCLUDE ALTERNATIVES)

6. WEATHER FORECAST FOR OPERATIONAL PERIOD

7. GENERAL/SAFETY MESSAGE

8. ATTACHMENTS (✓ IF ATTACHED)

☐ ORGANIZATION LIST (ICS 203)   ☐ MEDICAL PLAN (ICS 206)   ☐ _____

☐ DIVISION ASSIGNMENT LISTS (ICS 204)   ☐ INCIDENT MAP   ☐ _____

☐ COMMUNICATIONS PLAN (ICS 205)   ☐ TRAFFIC PLAN   ☐ _____

ICS 202	9. PREPARED BY (PLANNING SECTION CHIEF)	10. APPROVED BY (INCIDENT COMMANDER)

FORM 16-2   Incident Objectives
Source: NIIMS, ICS 202

ORGANIZATION ASSIGNMENT LIST (ICS 203)	1. INCIDENT NAME	2. DATE PREPARED	3. TIME PREPARED

POSITION	NAME	4. OPERATIONAL PERIOD (DATE/TIME)

5. INCIDENT COMMANDER AND STAFF

INCIDENT COMMANDER		**9. OPERATIONS SECTION**	
DEPUTY		CHIEF	
SAFETY OFFICER		DEPUTY	
INFORMATION OFFICER		**a. BRANCH I – DIVISIONS/GROUPS**	
LIAISON OFFICER		BRANCH DIRECTOR	

6. AGENCY REPRESENTATION

AGENCY	NAME

		DEPUTY	
		DIVISION/GROUP	
		DIVISION/GROUP	
		DIVISION/GROUP	
		DIVISION/GROUP	
		DIVISION/GROUP	
		b. BRANCH II – DIVISIONS/GROUPS	
		BRANCH DIRECTOR	

7. PLANNING SECTION

CHIEF		DEPUTY	
DEPUTY		DIVISION/GROUP	
RESOURCES UNIT		DIVISION/GROUP	
SITUATION UNIT		DIVISION/GROUP	
DOCUMENTATION UNIT		DIVISION/GROUP	
DEMOBILIZATION UNIT		DIVISION/GROUP	
TECHNICAL SPECIALISTS		**c. BRANCH III – DIVISIONS/GROUPS**	
		BRANCH DIRECTOR	
		DEPUTY	
		DIVISION/GROUP	
		DIVISION/GROUP	
		DIVISION/GROUP	
		DIVISION/GROUP	

8. LOGISTICS SECTION

CHIEF		DIVISION/GROUP	
DEPUTY		**d. AIR OPERATIONS BRANCH**	
a. SUPPORT BRANCH		AIR OPERATIONS BR. DIR.	
DIRECTOR		AIR ATTACK SUPERVISOR	
SUPPLY UNIT		AIR SUPPORT SUPERVISOR	
FACILITIES UNIT		HELICOPTER COORDINATOR	
GROUND SUPPORT UNIT		AIR TANKER COORDINATOR	
b. SERVICE BRANCH		**10. FINANCE SECTION**	
DIRECTOR		CHIEF	
		DEPUTY	
		TIME UNIT	
COMMUNICATIONS UNIT		PROCUREMENT UNIT	
MEDICAL UNIT		COMPENSATION/CLAIMS UNIT	
FOOD UNIT		COST UNIT	

ICS 203	PREPARED BY (RESOURCES UNIT)

FORM 16-3 Organization Assignment List
Source: NIIMS, ICS 203

1. BRANCH	2. DIVISION/GROUP	**DIVISION ASSIGNMENT LIST** (ICS 204)		

3. INCIDENT NAME	4. OPERATIONAL PERIOD DATE _____ TIME _____

5. OPERATIONS PERSONNEL

OPERATIONS CHIEF _____ DIVISION/GROUP SUPERVISOR _____

BRANCH DIRECTOR _____ AIR ATTACK SUPERVISOR _____

6. RESOURCES ASSIGNED THIS PERIOD

STRIKE TEAM/TASK FORCE/ RESOURCE DESIGNATOR	LEADER	NUMBER PERSONS	TRANS. NEEDED	DROP OFF PT./TIME	PICK UP PT./TIME

7. CONTROL OPERATIONS

8. SPECIAL INSTRUCTIONS

9. DIVISION/GROUP COMMUNICATION SUMMARY

FUNCTION		FREQ.	SYSTEM	CHAN.	FUNCTION		FREQ.	SYSTEM	CHAN.
COMMAND	LOCAL REPEAT				SUPPORT	LOCAL REPEAT			
DIV./GROUP TACTICAL					GROUND TO AIR				

PREPARED BY (RESOURCE UNIT LDR.)	APPROVED BY (PLANNING SECT. CH.)	DATE	TIME

FORM 16-4   Division Assignment List
Source: NIIMS, ICS 204

COMMUNICATIONS PLAN (ICS 205)	1. INCIDENT NAME	2. DATE/TIME PREPARED	3. OPERATIONAL PERIOD TO

4. COMMUNICATIONS RESOURCE ALLOCATION					
SYSTEM/CACHE	TALK GROUP/ FREQUENCY	FUNCTION	KNOW POSITION CHANNEL/PHONE #	ASSIGNMENT	REMARKS

PREPARED BY (COMMUNICATIONS UNIT LEADER)

FORM 16-5   Communications Plan
Source: NIIMS, ICS 205

MEDICAL PLAN (ICS 206)	1. INCIDENT NAME	2. DATE PREPARED	3. TIME PREPARED	4. OPERATIONAL PERIOD

5. INCIDENT MEDICAL AID STATIONS

MEDICAL AID STATIONS	LOCATION	PARAMEDICS	
		YES	NO

6. TRANSPORTATION

A. AMBULANCE SERVICES

NAME	ADDRES	PHON	PARAMEDICS	
			YES	NO

B. INCIDENT

NAME	LOCATION	PARAMEDICS	
		YES	NO

7. HOSPITALS

NAME	ADDRESS	TRAVEL TIME		PHONE	HELIPAD		BURN CENTER	
		YES	NO		YES	NO	YES	NO

8. MEDICAL EMERGENCY PROCEDURES

ICS 206	9. PREPARED BY (MEDICAL UNIT LEADER)	10. REVIEWED BY (SAFETY OFFICER)

FORM 16-6 Medical Plan

Source: NIIMS, ICS 206

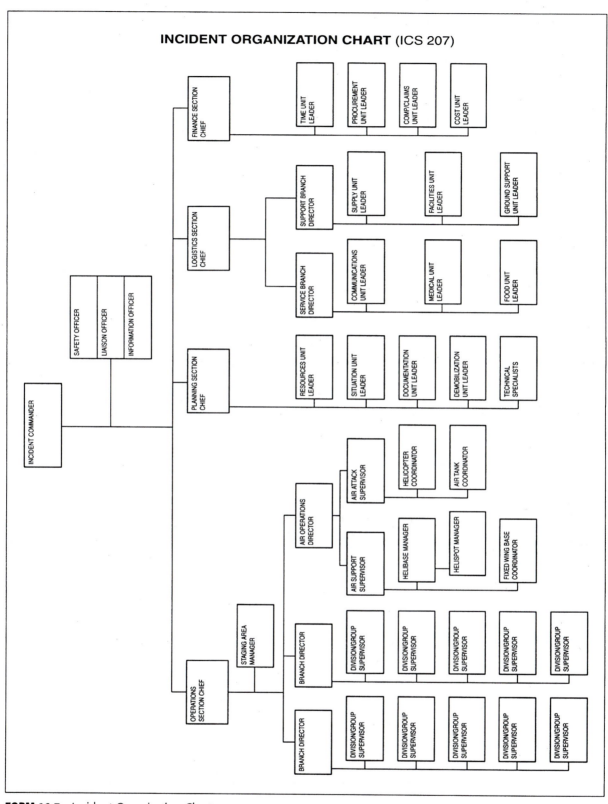

FORM 16-7 Incident Organization Chart

Source: NIIMS, ICS 207

1. INCIDENT NAME	2. INCIDENT NO.	3. INCIDENT COMMANDER	4. JURISDICTION	5. COUNTY	**INCIDENT STATUS SUMMARY** (ICS 209)

6. TYPE INCIDENT	7. LOCATION	8. STARTED (DATE/TIME)

9. CAUSE	10. AREA INVOLVED	11. PERCENT CONTAINED	12. EXPECTED CONTAINMENT Date _____ Time _____	13. PERCENT CONTROLLED	14. EXPECTED CONTROL Date _____ Time _____

15. CURRENT THREAT	16. CONTROL PROBLEMS

17. EST. LOSS	18. EST. SAVINGS	19. INJURIES _____ DEATHS _____	20. LINE BUILT	21. LINE TO BUILD

22. CURRENT WEATHER WS _____ TEMP _____ WD _____ RH _____	23. PREDICTED WEATHER NEXT PERIOD WS _____ TEMP _____ WD _____ RH _____	24. INCIDENT COSTS— PREVIOUS DAY	25. TOTAL COST TO DATE

26. AGENCIES / 27. RESOURCES KIND OF RESOURCES	INC	ST	INC	ST	INC	ST	INC	ST	INC	ST	INC	ST	INC	ST	INC	ST	INC	ST	INC	ST	INC	ST	TOTALS INC	ST
ENGINES																								
DOZERS																								
CREWS																								
HELICOPTERS																								
AIR TANKERS																								
TRUCK COS.																								
RESCUE/MED.																								
WATER TENDERS																								
OVERHEAD PERSONNEL																								
TOTAL PERSONNEL																								

28. COOPERATING AGENCIES

29. REMARKS

30. PREPARED BY	31. APPROVED BY	32. DATE _____ TIME _____	33. INITIAL ☐ UPDATE ☐ FINAL ☐	34. SENT TO DATE _____ TIME _____ BY _____

FORM 16-8   Incident Status Summary
Source: NIIMS, ICS 209

OPERATIONAL PLANNING WORKSHEET (ICS

Field	Label
1.	INCIDENT
2.	DATE / TIME
3.	OPERATIONAL PERIOD

6. RESOURCES BY TYPE (SHOWN STRIKE TEAM AS)

5. WORK ASSIGNMENTS		ENGINES (1 2 3 4)	WATER	HAND CREWS	DOZERS (1 2 3)	HELICOPTERS (1 2 3 4)	AIR TANKERS (1 2 3)	OTHE	7. REPORTING LOCATION	8. REQUESTED ARRIVAL
	REQ.									
	HAVE									
	NEED									
	REQ.									
	HAVE									
	NEED									
	REQ.									
	HAVE									
	NEED									
	REQ.									
	HAVE									
	NEED									
	REQ.									
	HAVE									
	NEED									
	REQ.									
	HAVE									
	NEED									
	REQ.									
	HAVE									
	NEED									
	REQ.									
	HAVE									
	NEED									

4. DIVISION/ GROUP OR OTHER LOCATION

9. TOTAL RESOURCES — SINGLE RESOURC — STRIKE TEAMS

TOTAL RESOURCES ON

TOTAL RESOURCES

PREPARED BY (NAME AND

ICS

FORM 16-9 Operational Planning Worksheet
Source: NIIMS, ICS 215

□ 2 + 1 MEDICAL   **EMS MEDICAL WORKSHEET**
    □ FIRST ALARM MEDICAL   INCIDENT NO. _____

ADDRESS _____ TIME _____

INCIDENT TYPE _____

① ② ③ ④ ⑤ ⑥ ⑦ □

□ INITIAL REPORT
□ INITIATE COMMAND
□ PERSONNEL PROTECTION
□ TRIAGE
□ EXTRICATION   □ ALL CLEAR
□ TREATMENT
□ TRANSPORTATION
□ TRAFFIC/CROWD CONTROL
□ HOSPITAL NOTIFICATION
□ SCENE STABILIZED
□ PROGRESS REPORTS

□ _____

COMMAND

FIRE DEPT		
EP		
EP		
EP		
E		
E		
L		
L		
R		
R		
BC		

EXTRICATION	TREATMENT	TRANSPORTATION	

RESCUE		
1		
2		
3		
4		

HELICOPTER			PRIORITY 1	PRIORITY 2	PRIORITY 3	PRIORITY 4
1						
2						

PT # _____ PRIORITY _____
UNIT(S) TREATING _____
UNIT TRANSPORTING _____
HOSPITAL _____
INJURY _____
NAME _____
ADDRESS _____
SEX _____ AGE _____

PT # _____ PRIORITY _____
UNIT(S) TREATING _____
UNIT TRANSPORTING _____
HOSPITAL _____
INJURY _____
NAME _____
ADDRESS _____
SEX _____ AGE _____

PT # _____ PRIORITY _____
UNIT(S) TREATING _____
UNIT TRANSPORTING _____
HOSPITAL _____
INJURY _____
NAME _____
ADDRESS _____
SEX _____ AGE _____

PT # _____ PRIORITY _____
UNIT(S) TREATING _____
UNIT TRANSPORTING _____
HOSPITAL _____
INJURY _____
NAME _____
ADDRESS _____
SEX _____ AGE _____

PT # _____ PRIORITY _____
UNIT(S) TREATING _____
UNIT TRANSPORTING _____
HOSPITAL _____
INJURY _____
NAME _____
ADDRESS _____
SEX _____ AGE _____

PT # _____ PRIORITY _____
UNIT(S) TREATING _____
UNIT TRANSPORTING _____
HOSPITAL _____
INJURY _____
NAME _____
ADDRESS _____
SEX _____ AGE _____

92-45.2D Rev. 4/94
61582505824-CP

FORM 16-10 EMS Medical Worksheet

Source: Phoenix Fire Department, Phoenix, AZ

TACTICAL WORKSHEET

Address: _____   Incident No. _____   Dispatch Time _____

Occupancy: _____

Wind Direction

_____

Radio Channel _____

Elapsed Time

5 10 15 20 25 30 PAR

Level II Staging Loc.

_____

Personnel Accountability
(PAR)

All Clear

Under Control

Off-To-Def

30 Min.

Hazardous Event

Missing Fire Fighter

No "PAR" Upgrade Assign.

Benchmarks

Tactical
- ☐ Strategy / Offense-Defense
- ☐ Action Plan
- ☐ Fire Attack
- ☐ Search & Rescue
- ☐ **All Clear (PAR)**
- ☐ **Fire Control (PAR)**
- ☐ Water Supply
- ☐ Pumped Water
- ☐ Property Conservation
- ☐ **Loss Stopped**
- ☐ Ventilation
- ☐ Exposures
- ☐ Rapid Intervention Crew (RIC)

Functional
- ☐ Accountability
- ☐ Command Location
- ☐ Evacuation
- ☐ Logistics
- ☐ Gas
- ☐ Electrical
- ☐ Recon
- ☐ Investigator
- ☐ Police
- ☐ Outside Agency
- ☐ Occupant Services
- ☐ Red Cross
- ☐ **Secondary All Clear**

Command

Branch   Branch

92-121D Rev. 10/96
61582506681-CP

FORM 16-11 Tactical Worksheet
Source: Phoenix Fire Department, Phoenix, AZ

POST-INCIDENT REVIEW WORKSHEET

Units / Disp.	From Quarters?	Turnouts / When?	Account. / Locations	Staged Per S.O.P.s / Assignment	Inc. #_____ Date _____

_____ _____ _____ _____ _____

_____ _____ _____ _____ _____

_____ _____ _____ _____ _____

_____ _____ _____ _____ _____

_____ _____ _____ _____ _____

_____ _____ _____ _____ _____

_____ _____ _____ _____ _____

_____ _____ _____ _____ _____

Strategy (Offensive or Defensive) (Marginal* *rescue only*)   **RIC Team** _____

Initial _____ 10 min. _____ 20 min. _____ 30 min. _____

Command Mode (Command-Fast Attack-Nothing Showing)? _____ Crew Assign. _____

On-scene Report: _____

Incident Command Roll Call

Initial IC _____ IC#2 _____ IC#3 _____ IC#4 _____ SO _____ SA _____

Incident Action Plan: _____

Water Supply: Initial (1st Eng.?) _____ 2nd Supply _____ Pumped _____

Benchmarks: All Clear *(PAR)* _____ Fire Control *(PAR)* _____ Loss Stopped_____ 30 min *(PAR)* _____

Progress Reports: (yes or no) _____ (Change Plan?) _____

Ventilation: Who? _____ PPV _____ Vertical _____ Other _____   **Utilities Secured:** _____

Reason? _____

Effectiveness _____

Property Conservation: Who? _____ Occ. Serv. _____ Salvage _____ Overhaul _____ Red Cross _____

Customer Response _____

Structural/Fire Evaluation: _____

Rehabilitation/RIC Activities: _____

Accidents, Injuries/Near Misses: _____

Comments: _____

_____

_____

_____

_____

Fire Fighter Safety, Loss Control, and Customer Service are ongoing throughout the incident.

92-121D Rev 10/96
61582506681-CP

FORM 16-12   Post-Incident Review Worksheet

Source: Phoenix Fire Department, Phoenix, AZ

17

Health, Fitness, and Wellness Resource Directory

Laurence J. Stewart

In this health, fitness, and wellness resource directory, the phone numbers, addresses, and website information for the various organizations have been updated. The resource list has also been expanded to include additional areas of concern in the field of health, fitness, and wellness.

National Health Information Center (NHIC)

P.O. Box 1133
Washington, DC 20013-1133
1-800-336-4797
www.healthfinder.gov

AIDS

AIDS Information Hotline
U.S. Public Health Service
American Social Health Association

P.O. Box 13827
Research Triangle Park, NC 27709
1-800-342-AIDS (2437)
1-800-344-7432 (in Spanish)
1-800-243-7889 (for hearing impaired)
www.cdc.gov/hiv

AIDS Treatment Information Services

1-800-HIV-0440 (1-800-448-0440)

CDC National Prevention Information Network

HIV/AIDS/TB/STD's
www.cdcnpin.org

Project Inform HIV/AIDS Treatment Hotline

1-800-822-7422
www.projinf.org

ALCOHOL/DRUG ABUSE

Al-Anon Family Group Headquarters (includes Alateen)

P.O. Box 862
Midtown Station
New York, NY 10018-0862
1-800-356-9996
alanon.alateen.org

Laurence J. Stewart, an associate fire service specialist for the NFPA Public Fire Protection Division, has 20 years experience within the fire and emergency services.

Alcoholics Anonymous (AA)
General Service Office

475 Riverside Drive
New York, NY 10115
212-870-3400
To find an AA group in your area, call:
Intergroup 212-647-1680
www.alcoholics-anonymous.org

The American Council on Alcoholism

1-800-527-5344
http://www.aca-usa.org/

The National Council on Alcoholism
and Drug Dependency

1-800-622-2255
http://www.ncadd.org/

Children of Alcoholics Foundation
(COAF)

555 Madison Ave. 20th Floor
New York, NY 10022
1-800-359-COAF (2623)
www.social.com

Cocaine Anonymous World Services, Inc.

3740 Overland Ave., Suite G
Los Angeles, CA 90034
1-800-347-8998
213-554-2554
www.recovery.netwiz.net/ca

Men for Sobriety (MFS) and Women
for Sobriety (WFS) Inc.

P.O. Box 618
Quakertown, PA 18951-0618
1-800-333-1606
www.womenforsobriety.org

Narcotics Anonymous (NA)
NA World Services

P.O. Box 9999
Van Nuys, CA 91409
1-818-773-9999
www.wsoinc.com

National Cocaine Hotline
c/o Phoenix House

164 West 74th St.
New York, NY 10023
1-800-COCAINE (262-2463)
1-800-662-HELP (4357)
www.drughelp.org

Alcohol and Drug Abuse Helpline

107 Lincoln St.
Worcester, MA 01605
1-800-ALCOHOL (662-4357)
Drug-Free Workplace Hotline:
1-800-843-4571
www.adcare.com

Rational Recovery (RR)
Rational Recovery Self-Help Network
(RRSN)

Box 800
Lotus, CA 95651
916-621-2667
www.rational.org/recovery

ALZHEIMER'S DISEASE

Alzheimer's Association

919 North Michigan Ave., Suite 1000
Chicago, IL 60601-1676
1-800-272-3900
www.alz.org

Alzheimer's Disease Education
and Referral Center

1-800-438-4380
www.alzheimers.org

ANXIETY/PHOBIAS

Anxiety Disorders Association
of America (ADAA)

11900 Parklawn Drive, Suite 100
Rockville, MD 20852
301-231-9350
www.adaa.org

ARTHRITIS

Arthritis Foundation

1330 West Peachtree Street
Atlanta, GA 30309
1-800-238-7000
www.arthritis.org

ADD/ADHA

Attention Deficit Disorder Association

1-800-487-2282
www.add.org

CHADD-Children & Adults with Attention Deficit/Hyperactivity Disorder

1-800-233-4050
www.chadd.org

CANCER

Cancer Information Service (National Cancer Institute)

9000 Rockville Pike
Bethesda, MD 20892
1-800-4-CANCER (422-6237)
www.nci.nih.gov

American Cancer Society Response Line

1-800-227-2345
www.cancer.org

Y-ME National Breast Cancer Organization

1-800-221-2141
www.y-me.org

CHILD ABUSE/MISSING CHILDREN

Boys Town National Hotline

1-800-448-3000
www.boystown.org

ChildHelp/IOF Foresters National Child Abuse Hotline

1-800-422-4453
www.iof.org

National Center for Missing and Exploited Children Hotline

1-800-843-5678
www.ncmec.org

National Clearinghouse on Child Abuse and Neglect

1-800-394-3366
www.calib.com

National Council on Child Abuse and Family Violence

1-800-222-2000
www.nccafv.org

National Runaway Switchboard

1-800-621-4000
www.nrscrisisline.org

DEPRESSION

National Depressive and Manic Depression Association

730 North Franklin, Suite 501
Chicago, IL 60610
312-642-0049
www.ndmda.org

DIABETES

American Diabetes Association

1660 Duke St.
Alexandria, VA 22314
1-800-232-3472
www.diabetes.org

Juvenile Diabetes Foundation International

1660 Duke St.
Alexandria, VA 22314
1-800-232-3472
www.diabetes.org

DISABILITIES

American Rehabilitation Association

1-800-368-3513
www.amrpa.org

National Easter Seal Society

1-800-221-6827
www.easter-seals.org

National Information Center for Children and Youth with Disabilities

P.O. Box 1492
Washington, DC 20013-1492
1-800-695-0285
www.nichcy.org

National Clearinghouse for Children Deaf and Blind

1-800-438-9376

DOMESTIC VIOLENCE

National Child Abuse Hotline Child Help U.S.A.

15757 North 78th Street
Scottsdale, AZ 85260
1-800-4-A-CHILD (422-4453)
www.childhelpusa.org

National Domestic Violence Hotline

1-800-799-SAFE (7233)

National Resource Center on Domestic Violence

1-800-537-2238

EATING DISORDERS

National Association of Anorexia Nervosa and Associated Disorders (ANAD)

P.O. Box 7
Highland Park, IL 60035
847-831-3438
www.anad.org

National Eating Disorder Hotline and Referral Service

847-831-3438

Food Addiction Hotline

1-800-248-3285

FITNESS

Aerobics and Fitness Association of America

1-800-233-4886
www.afaa.com

American College of Sports Medicine

1-800-486-5643
www.acsm.org

GRIEF

The Compassionate Friends National Support Group

900 Jorie Blvd.
Oakbrook, IL 60521
708-990-0010
www.compassionatefriends.org

Grief Recovery Institute

www.grief-recovery.com

Grief Recovery Helpline

1-800-445-4808

HEADACHES

Brain Injury Association

1-800-444-6443
www.biausa.org

National Headache Foundation

428 W. St. James Place
Chicago, IL 60614
1-800-843-2256
www.headaches.org

HEARING AND SPEECH PROBLEMS

American Speech-Language Hearing Association

1-800-638-8255
www.asha.org

Better Hearing Institute—Hearing Help Line

P.O. Box 1840
Washington, DC 20013
1-800-327-9355
www.betterhearing.org

Dial A Hearing Screening Test

1-800-222-3277

Hear Now

1-800-648-4327

Hearing Aid Helpline

1-800-521-5247

John Tracy Clinic

1-800-522-4582
www.johntracyclinic.org

National Center for Stuttering

1-800-221-2483
www.stuttering.com

Stuttering Foundation of America

1-800-992-9392
www.stuttersfa.org

The LEAD LINE

1-800-352-8888

HEART DISEASE/HIGH BLOOD PRESSURE

American Heart Association

7272 Greenville Ave.
Dallas, TX 75231-4596
1-800-242-8721
www.americanheart.org

National Heart Lung and Blood Institute (NHLBI) Information Center

P.O. Box 30105
Bethesda, MD 20824-0105
1-800-575-WELL (9355)
301251-1222
www.nhlbi.nih.gov/index.htm

HOSPITAL/HOSPICE CARE

Hill-Burton Hospital Free Care

1-800-638-0742

Hospice Education Institute

"Hospice Link"
1-800-331-1620
www.hospiceworld.org

IMPOTENCE

The Gedding Osbon Foundation Impotence Resource Center

1-800-433-4215
www.impotence.org

KIDNEY DISEASES

American Kidney Fund

6110 Executive Blvd., Suite 1010
Rockville, MD 20852
1-800-638-8299
www.arebon.com/kidney

LEARNING DISORDERS

The Orton Dyslexia Society

1-800-222-3123

LIVER DISEASES

American Liver Foundation

1425 Pompton Avenue
Cedar Grove, NJ 07009
1-800-223-0179
www.liverfoundation.org

LUNG DISEASES

American Lung Association

1740 Broadway
NY, NY 10019
212-315-8700
www.lungusa.org

Asthma and Allergy Foundation of America

1233 20th Street, NW, Suite 402
Washington, D.C. 20036
1-800-727-8462
www.aafa.org

Asthma Information Center

1-800-727-5400

National Jewish and Medical Research Center - Lung Line

1400 Jackson St.
Denver, CO 80206
1-800-222-LUNG
www.njc.org

MEDICAL IDENTIFICATION

MedicAlert Foundation International

P.O. Box 1009
Turlock, CA 95381-1009
1-800-344-3226
www.medicalert.org

MENTAL HEALTH

National Clearinghouse on Family Support and Children's Mental Health

1-800-628-1696

National Foundation for Depressive Illness

P.O. Box 2257
New York, NY 10116
1-800-239-1265
www.depression.org

National Institute of Mental Health Information Line

1-800-421-4211 and 1-800-647-2642
www.nimh.nih.gov

National Mental Health Association

1021 Prince Street
Alexandria, VA 22314-2971
1-800-969-6642
www.nmha.org

NUTRITION

National Center for Nutrition and Dietetics
American Dietetic Association

216 West Jackson Blvd., Suite 800
Chicago, IL 60606
Consumer Nutrition Information Hotline
1-800-366-1655
www.eatright.org

ORGAN DONATION

The Living Bank

1-800-528-2971
www.livingbank.org

United Network for Organ Sharing

1-800-243-6667
www.unos.org

OSTEOPOROSIS

National Osteoporosis Foundation

1150 17th Street N.W., Suite 500
Washington, D.C. 20036
202-223-2226
www.nof.org

Osteoporosis and Related Bone Diseases
National Resource Center

1150 17th Street, NW, Suite 500
Washington, DC 20036
1-800-624-BONE (2663)
www.osteo.org

PARALYSIS AND SPINAL CORD INJURY

American Paralysis Association

500 Morris Avenue
Springfield, NJ 07081
1-800-225-0292
www.apacure.com

APA Spinal Cord Injury Hotline

1-800-526-3456

PARKINSON'S DISEASE

National Parkinson Foundation

1501 N.W. 9th Ave.
Bob Hope Rd.
Miami, FL 33136
1-800-327-4545
www.parkinson.org

PREGNANCY/MISCARRIAGE

Bradley Method of Natural Childbirth

1-800-423-2397
www.naturalbirth.org

ASPO/Lamaze

1-800-368-4404
www.lamaze-childbirth.com

AAO National Problem Pregnancy Hotline

1-800-228-0332

RARE DISORDERS

American SIDS Institute

2480 Windy Hill Road, Suite 380
Marietta GA 30067
www.sids.org

Cleft Palate Foundation

104 South Estes Drive, Suite 204
Chapel Hill, NC 27514
919-933-9044
1-800-242-5338
www.cleftline.org

Crohn's and Colitis Foundation of America

1-800-932-2423
www.ccfa.org

Cystic Fibrosis Foundation

6931 Arlington Road
Bethesda, Maryland 20814
1-800-344-4823
www.cff.org

Epilepsy Foundation of America

4351 Garden City Dr., 4th Floor
Landover, MD 20785
1-800-332-1000
www.efa.org

Lupus Foundation of America

1-800-558-0121
www.lupus.org

National Association for Sickle Cell Disease

200 Corporate Pointe, Suite 495
Culver City, California 90230
1-800-421-8453
www.sicklecelldisease.org

National Down Syndrome Congress

7000 Peachtree-Dunwoody Road, N.E.
Lake Ridge 400 Office Park Building 5,
Suite 100
Atlanta, GA 30328-1655
www.ndsccenter.org

National Down Syndrome Society

666 Broadway
New York, NY 10012
1-800-221-4602
www.ndss.org

National Multiple Sclerosis Society

733 Third Ave.
New York, NY 10017-3288
1-800-344-4867
www.nmss.org

National Organization for Rare Disorders

P.O. Box 8923
New Fairfield, CT 06812-8923
1-800-999-6673
www.rarediseases.org

National Reye's Syndrome Foundation

P.O. Box 829
Bryan, OH 43506
1-800-233-7393
www.bright.net/~reyessyn

United Cerebral Palsy Association

1660 L Street, NW, Suite 700
Washington, DC 20036
1-800-872-5827
www.ucpa.org

SAFETY

National Lead Information Hotline

1-800-532-3394

National Safety Council

1121 Spring Lake Drive
Itasca, Illinois 60143-3201
1-800-621-7619
www.nsc.org

U.S. Consumer Product Safety Commission Hotline

1-800-638-2772

U.S. DOT Auto Safety Hotline

1-800-424-9393

SENIOR CITIZEN HEALTH

American Association of Retired Persons (AARP)

601 E St. N.W.
Washington, DC 20049
1-800-424-2277
www.aarp.org

ElderCare Locator

1-800-677-1116

National Institute of Aging (NIA) Information Center

P.O. Box 8057
Gaithersburg, MD 20898-8057
1-800-222-2225
www.nih.gov/nia/

SEXUAL EDUCATION

Planned Parenthood Federation of America, Inc.

1-800-230-7526
www.plannedparenthood.org

SEXUALLY TRANSMITTED DISEASES

Centers for Disease Control

National STD Hotline
1-800-227-8922

STRESS

The American Academy of Experts in Traumatic Stress

631-543-2217
www.aaets.org

STROKE

American Heart Association Stroke Connection

1-800-553-6321
www.strokeassociation.org

National Stroke Association

9707 E. Easter Lane
Englewood, Co. 80112
1-800-787-6537
www.stroke.org

SUICIDE/SELF-ABUSE

American Suicide Foundation

120 Wall Street, 22nd Floor
New York, New York 10005
1-800-531-4477 (For information. Not a crisis hotline)
www.afsp.org

Self-Injury Hotline

SAFE (Self-Abuse Finally Ends) Alternative Programs
1-800-366-8288
www.selfinjury.com

Suicide & Crisis Hotline

1-800-999-9999

Yellow Ribbon Suicide Prevention Program

1-800-SUICIDE (784-2433)
www.yellowribbon.org

UROLOGICAL DISORDERS

American Association of Kidney Patients

3505 E. Frontage Rd., Ste. 315
Tampa, FL 33607
1-800-749-2257
www.aakp.org

National Kidney Foundation

1-800-622-9010
www.kidney.org

The Simon Foundation for Continence

1-800-237-4666
www.simonfoundation.org

VISUAL PROBLEMS

American Council for the Blind

1155 15th Street N.W., Suite 720
Washington, DC 20005
1-800-424-8666
www.acb.org

Blind Children's Center

1-800-222-3566
www.blindkids.org

National Association for Parents of the Visually Impaired

1-800-562-6265
www.spedex.com/NAPVI

National Federation of the Blind

1800 Johnson Street
Baltimore, MD 21230
410-659-9314
www.nfb.org

Prevent Blindness America

500 E. Remington Rd.
Schaumburg, IL 60173
1-800-221-3004
www.prevent-blindness.org

WOMEN

Endometriosis Association

8585 North 76th Place
Milwaukee, WI 53223
1-800-992-3636
www.endometriosisassn.org

Women's Health America Group PMS Access/Menopause Natural Hormone Hotline

1289 Deming Way
Madison, WI 53717
1-800-222-4767
www.womenshealth.com

18

Guide for Fire Department Administrators*

The medical qualifications involved when a prospective fire department candidate is offered a position or for the continued employment of a current member entail a myriad of issues. With the changing technology in medical science, the issues that were relevant in the first edition of NFPA 1582, *Standard on Medical Requirements for Fire Fighters and Information for Fire Department Physicians,* may no longer be valid.

The fire department administrator may not be qualified to make a determination whether department personnel are medically qualified. The information in this annex material is provided as a guide to assist those who are attempting to use the standard as part of an overall fire department occupational safety and health program. Although some may view this standard as punitive, it is provided to assist a fire department in keeping its personnel well during their careers and after they leave the fire department. Just as the medical technology continually changes, so does the law. Users are advised to contact their municipal counsel for updates on laws cited in this annex material.

B.1 Legal Considerations in Applying the Standard.

The consideration of an application or continued employment of a member based on medical or physical performance evaluations involves a determination that is not without legal implications. To this end, prior to making an adverse employment decision based on the current standard, the authority with jurisdiction may wish to consult with legal counsel.

B.1.1 Legal Protections for Individuals with Handicaps or Disabilities.

The Rehabilitation Act of 1973, as amended, 29 U.S.C. § 791 et seq., and implementing regulations, prohibit discrimination against those with handicaps or disabilities under any program receiving financial assistance from the federal government. The Americans with Disabilities Act (ADA) of 1990, 42 U.S.C. § 12101, et seq., also prohibits employment discrimination by certain private employers against individuals with disabilities. In addition, many

---

*Editor's Note: This chapter appears as an annex to the 2003 edition of NFPA 1582, *Standard on Medical Requirements for Fire Fighters and Information for Fire Department Physicians.* It is not a part of the requirements of NFPA 1582 but is included with that document for informational purposes only.

states have enacted legislation prohibiting discrimination against those with handicaps or disabilities. Generally speaking, these laws prevent the exclusion, denial of benefits, refusal to hire or promote, or other discriminatory conduct against an individual based on a handicap or disability, where the individual involved can, with or without reasonable accommodation, perform the essential functions of the job without creating undue hardship on the employer or program involved.

Beginning in 1999, the United States Supreme Court has issued a series of decisions limiting the scope of the ADA. As a result, persons with certain kinds of impairments that are mitigated by corrective measures such as medication for high blood pressure or eyeglasses for myopia are not "disabled" under the ADA. See *Sutton v. United Airlines, Inc.*, 527 U.S. 471 (1999); *Murphy v. United Parcel Service, Inc.*, 118 S.Ct 2133 (1999); *Albertsons, Inc. v. Kirkingburg*, 527 U.S. 555 (1999). More recently the Supreme Court held that an impairment is not a disability covered by the ADA unless it severely restricts a person from doing activities that are of central importance to most peoples daily lives. *Toyota Motor Mfgr., Kentucky, Inc. v. Williams*, __U.S.__ (2002). These cases significantly limit the persons who can claim the protections of the federal ADA, but do not, by any means, eliminate the ADA as an important consideration in fire service-related employment decisions. Moreover, it should be borne in mind that separate disability protections exist under laws of many states, and some of these laws have been interpreted to afford greater protections than that afforded by the ADA. See, for example, *Dahill v. Boston Department of Police*, 434 Mass. 233 (2001), where the Supreme Court of Massachusetts ruled that a corrective device to alleviate a disability is not relevant in determining whether someone is disabled under the state's disability law.

The disability discrimination laws, therefore, continue to be an important part of the legal framework that governs employment-related decisions. Although this standard has been developed with this in mind, these laws can, depending on the jurisdiction and the circumstances, affect the degree to which the authority having jurisdiction can implement the standard in an individual case. Users of this standard should be aware that, while courts, in assessing disability discrimination claims, are likely to give considerable weight to the provisions of a nationally recognized standard such as NFPA 1582 [see, for example, *Miller v. Sioux Gateway Fire Department*, 497 N.W.2d 838 (1993)]. Reliance on the standard alone may not be sufficient to withstand a challenge to an adverse employment decision.

B.1.2 Legal Protections for Individuals Who Are Members of Protected Classes (Race, Sex, Color, Religion, or National Origin).

Title VII of the Civil Rights Act of 1964, as amended, 42 U.S.C. § 2000e, and implementing regulations by the Equal Employment Opportunity Commission (EEOC), prohibit discrimination in employment on the basis of race, sex, color, religion, or national origin (i.e., protected classes). Under Title VII, an "employer" is defined, generally, to mean a person with "15 or more employees for each working day in each of 20 or more calendar weeks in the current or preceding calendar year." (42 U.S.C. § 2000e) Several federal jurisdictions have held that unpaid volunteers are not considered to be "employees" under Title VII.

Additionally, many states, cities, and localities have adopted similar legislation. Generally, physical performance or other requirements that result in "adverse impact" on members of a protected class (e.g., on the basis of gender) are required to be validated through a study in accordance with EEOC guidelines, if such requirements are to be relied on in making employment decisions. Under EEOC guidelines, a study validating employment standards in one jurisdiction can be transportable to another jurisdiction (and therefore used in lieu of conducting a separate study). However, specific preconditions must be met in this regard, and the authority having jurisdiction should seek the advice of counsel before relying on a transported validation study.

B.1.2.1 Pregnancy and Reproduction.

Federal regulations, as well as many court decisions, including the U.S. Supreme Court's decision in *International Union, et al. v. Johnson Controls, Inc.* [499 U.S. 187, 111 S.Ct. 1196 (1991)], have interpreted the requirements of Title VII with respect to pregnancy and reproduction. The authority having jurisdiction should seek the advice of counsel in resolving specific questions concerning these requirements as well as other requirements that can be imposed by state or local laws.

B.2 Determining Essential Job Tasks.

The medical requirements in this edition of the standard were revised based on the essential job tasks contained in Chapter 5 and Chapter 9. It is recognized that some fire-fighting functions and tasks can vary from location to location due to differences in department size, functional and organizational differences, geography, level of urbanization, equipment utilized, and other factors. Therefore, it is the responsibility of each individual fire department to document, through job analysis, the essential job functions that are performed in the local jurisdiction.

There are a wide variety of job analytic techniques available to document the essential functions of the job of a member. However, at a minimum, any method utilized should be current, in writing, and meet the provisions of the Department of Labor regulations [29 CFR § 1630.2(n)(3)]. Job descriptions should focus on critical and important work behaviors and specific tasks and functions. The frequency and/or duration of task performance, and the consequences of failure to safely perform the task, should be specified. The working conditions and environmental hazards in which the work is performed should be described.

The job description should be made available to the fire department physician for use during the pre-placement medical examination for the individual determination of the medical suitability of applicants for member.

B.3 Choosing a Fire Department Physician.

Several factors should be considered in choosing a fire department physician. There are relatively few physicians with formal residency training and certification in occupational medicine. The fire department physician should be qualified to provide professional expertise in the areas of occupational safety and health as these areas relate to emergency services. For the purpose of conducting medical evaluations, the fire department physician should understand the physiological and psychological demands placed on members as well as the environmental conditions under which members have to perform.

Knowledge of occupational medicine and experience with occupational health programs, are essential for physicians not formally trained in occupational medicine.

The physician must be committed to meeting the requirements of the program, including appropriate record keeping. The physician's willingness to work with the department to continually improve the program is also important. Finally, the physician's concern and interest in the program and in the individuals are vital.

The following are some of the many options for obtaining physician services:

(1) Physicians may be paid on a service basis or through a contractual arrangement.

(2) For volunteer departments, local physicians may be willing to volunteer their services for the program, with other arrangements for payment of laboratory testing, X-rays, and so forth.

(3) Some departments may utilize a local health care facility for medical care. However, in that case, the department should have one individual physician responsible for the program, record keeping, and so forth.

(4) The use of a military reserve or a National Guard unit.

B.4 Coordinating the Medical Evaluation Program.

An individual from within the department should be assigned the responsibility for managing the health and fitness program, including the coordination and scheduling of evaluations and examinations. This person should also act as liaison between the department and the physician to make sure that each has the information necessary for decisions about placement, scheduling appointments, and so forth.

B.5 Confidentiality.

Confidentiality of all medical data is critical to the success of the program. Members need to feel assured that the information provided to the physician will not be inappropriately shared. No fire department supervisor or manager should have access to medical records without the express written consent of the member. There are occasions, however, when specific medical information is needed to make a decision about placement, return to work, and so forth, and a fire department manager should have more medical information for decision making. In that situation, written medical consent should be obtained from the member to release the specific information necessary for that decision.

Budgetary constraints can affect the medical program. Therefore, it is important that components of the program be prioritized such that essential elements are not lost. With additional funding, other programs or testing can be added to enhance the program.

19

Fire Service Joint Labor Management Wellness-Fitness Initiative*

The Fire Service Joint Labor Management Wellness-Fitness Initiative is a complete physical fitness and wellness program package. The developed program includes a manual, video, and data collection protocol. The participating fire chiefs and IAFF local union presidents have contributed to developing an overall wellness/fitness system with a holistic, positive, rehabilitating, and educational focus.

WHAT IS THE FIRE SERVICE JOINT LABOR MANAGEMENT WELLNESS-FITNESS INITIATIVE?

The Fire Service Joint Labor Management Wellness-Fitness Initiative is a historic partnership between the IAFF and the IAFC to improve the wellness of fire department uniformed personnel. Ten U.S. and Canadian public professional fire departments participated. All of these departments have committed themselves to this Wellness-Fitness Initiative by requiring the mandatory participation of all of their uniformed personnel in this program. This bold move to commit labor and management to the wellness of their uniformed personnel will carry the fire service into the 21st century.

The Fire Service Joint Labor Management Wellness-Fitness Initiative is intended to be implemented as a positive individualized program that is not punitive. All component results are measured against the individual's previous examinations and assessments and not against any standard or norm. However, medical practice standards may be used when results indicate that life saving intervention is required.

Confidentiality of medical information is a critical aspect of the Wellness-Fitness Initiative. The unauthorized release of personal details, which may be recorded as part of a medical evaluation, can and does cause legal, ethical, and personal problems either for the employee, the employer, or the examining physician. The Task Force

---

*This chapter, which is the introductory chapter from the manual of the Fire Service Joint Labor Management Wellness-Fitness Initiative, provides an overview of this fitness and wellness program.

Source: Excerpts from the Fire Service Joint Labor Management Wellness-Fitness Initiative reproduced by permission of the International Association of Fire Chiefs.

agrees that all information obtained from medical and physical evaluations is confidential, and the employer will only have access to information regarding fitness for duty, necessary work restrictions, and appropriate accommodations. The Task Force also agrees that all medical information must be maintained in separate files from all other personnel information.

What Is Wellness?

Wellness is a comprehensive term that includes all of the following:

- Medical Fitness
- Physical Fitness
- Emotional Fitness
- Access to rehabilitation, when indicated

Wellness programs in the fire department are intended to strengthen uniformed personnel so that their mental, physical, and emotional capabilities are resilient enough to withstand the stresses and strains of life and the workplace.

A wellness program is not just another program; it is a total commitment to:

- The health, safety, and longevity of all uniformed personnel
- The productivity and performance of all fire crews
- The cost effectiveness and welfare of all fire departments

Why Do Uniformed Personnel Need Wellness?

Each year, IAFF Death and Injury Surveys demonstrate that fire fighting remains one of the most dangerous occupations in the United States. Research has repeatedly shown the need for high levels of fitness to perform safely in the fire service.[1] The fire fighters' long hours, shift work, sporadic high intensity work, strong emotional involvement, and exposure to human suffering place fire fighting among the most stressful occupations in the world. [2, 3, 4] High levels of stress, intense physical demands, and long-term exposure to chemicals and infectious disease contribute to heart disease, lung disease, and cancer—the three leading causes of death and occupational disease disability.

Wellness is a personal commitment that all uniformed personnel must make to survive and to sustain a career in the professional fire service. When uniformed personnel are ill or injured, malnourished or overweight, overstressed or out of balance, it affects their ability to effectively do their job.

---

[1]Gledhill, N., and Jamnik, V. K. Characterization of the physical demands of fire fighting. *Can. J. Sports Sci.* 17 (1992): 297-313.

[2]Gist, R., and Woodall, S. J. Occupational stress in contemporary fire service: State of the art reviews. *Occupational Medicine* 10 (4): 763-777 (1995).

[3]Krantz, L. *Jobs Rated Almanac, Two.* Mahwah, NJ: World Almanac, 1992.

[4]IAFF. *Guide to Developing Fire Service Labor/Employee Assistance and Critical Incident Stress Management Programs,* 1999.

The benefits of wellness for uniformed personnel are many. They include: [5]

- Greater strength and stamina
- Weight reduction and/or control (maintenance)
- Lower cholesterol and blood pressure levels
- Decreased risk of death, injury, or disability from disease
- Heightened job performance and enjoyment from work
- Improved performance in physical activities
- Better posture and joint functioning
- Reduction of anxiety, stress, tension, and depression
- Increased energy, general vitality, and mental sharpness
- Enhanced self-esteem and self-image
- More restful and refreshing sleep
- Enhanced capacity to recover from strenuous and exhaustive work
- Increased tolerance for heat stress and more effective body cooling
- Improved mobility, balance, and coordination

Why Does the Union Need Wellness?

Fire fighter unions must assume a leadership role in implementing wellness/fitness programs for their members. Safety in fire fighting traditionally has meant purchasing the latest equipment available, such as new pumpers, bunker gear, or PASS alarms. Yet, the most important component in emergency response is the fire fighter and EMS provider. The definition for safety in fire fighting must expand to include a wellness/physical fitness program for the fire fighter. Unions must work to ensure that all members have the opportunity to attain and maintain a healthy body and mind so they can perform their work duties.

The responsibility for wellness/physical fitness programs cannot just be given to management. Without union input and cooperation in the process, members will not "buy in" to the program. Labor and management together must develop a wellness program that is educational and rehabilitative and not punitive.

Fire fighter unions work hard to improve the economic status of their members. A quality wellness/fitness program will help all members perform their duties and allow them to enjoy the fruits of their labor when they retire.

Why Does a Fire Chief Need Wellness?

As previously stated, wellness is a commitment that all uniformed personnel must make to survive the job. It is the fire chief's job to ensure that excellent customer service for the community is delivered by healthy fire fighters, with their performance enhanced by an atmosphere of workplace safety, regulatory compliance, and positive attitudes. Wellness is the fire chief's commitment to the uniformed personnel's quality of life. It is a commitment to the health of uniformed personnel when they come to work, respond to calls, return from calls, go home at the end of their shifts, and retire at the end of their careers. Wellness will facilitate compliance with workplace

---

[5]Pearson, J.; Hayford, J.; and Royer, W. *Comprehensive Wellness for Fire Fighters: Fitness and Health Guide for Fire and Rescue Workers*. New York: Van Nostrand Reinhold, 1995, p. 1.

regulations and improve the responsiveness of fire chiefs to directives from governing political bodies. Finally, in most departments, a fire chief is not only an administrator but also an active fire fighter, subject to the same stresses of heat, dehydration, noxious exposures, and other occupational hazards. Thus, a fire chief's commitment to wellness serves both personal and professional interests.

Why Should the Community Support Wellness?

Every fire incident or response within the community is unique. The ability of uniformed personnel to respond effectively is enhanced by their level of physical and mental preparedness. A wellness program is cost-effective for the community. Injury rates and sick leave usage can be reduced, thereby controlling overtime costs associated with filling vacant positions or utilizing other agencies for response. Wellness programs can facilitate fire department compliance with federal, state, and local laws related to issues such as infectious disease training and testing and breathing apparatus certification. The use of costly outside consulting agencies for these services would be reduced. Fire departments with members who are medically, physically, and mentally fit will provide better service to their communities year after year while realizing reductions in disability retirements by their uniformed personnel.

Financial and Administrative Commitment

The implementation of a wellness program is not free. There may be, however, significant cost benefits to initiating or expanding wellness programs. Wellness programs have repeatedly been shown to provide long-term savings. Many large corporations, including AT&T Communications, Union Pacific Railroad, DuPont Chemical Company, and The Travelers Corporation tout returns of $1.50 to $3.40 for every dollar invested in their wellness efforts.[6] A recent study addressing the effectiveness of a wellness program for an offshore oil exploration company demonstrated a threefold reduction in back injuries, as well as dramatic reductions in lost workday injuries (2.6-fold), non-lost workday injuries (12-fold), and first aid cases (2.3-fold), at a cost savings of $800,000 and a return on investment of $2.51.[7] The value of this investment has been shown in the fire service as well. In January 1997, the City of Phoenix, Arizona, conducted an audit of its disability retirement program for all city employees. The annualized cost of disability pensions for fire fighters was $100,000; for police officers, with twice as many personnel as the fire department, the annualized cost was $721,000; and, for general city employees, with five times as many personnel as the fire department, the annualized cost was $623,000. The reduced disability pension cost for the Phoenix Fire Department reflects its twelve-year commitment to an effective wellness program and thorough rehabilitation for all fire fighters.[8]

The fire service's greatest asset is not equipment, apparatus, or stations, but rather its personnel. It is through personnel that the fire departments serve the public, accomplish their missions, and are able to make a difference in their communities. By committing to a wellness program, the fire department gains the members' trust. This trust enhances every program and each call answered by the fire department. Therefore, placing a high priority on wellness makes sense for everyone, including fire service personnel, taxpayers, and the public served.

---

[6]National Institute of Consulting Services, 1996.

[7]Maniscalco, P.; Lane, R.; Welke, M.; Mitchell, J.; and Husting, L. Decreased rate of back injuries through a wellness program for offshore petroleum employees. *J. Occ. Env. Medicine* 41 (9): 813-820 (1999).

[8]City Auditor, City of Phoenix, AZ. *Disability Retirement Program Evaluation.* January 28, 1997.

Is Wellness Important to All Uniformed Personnel Regardless of Their Assigned Tasks?

Wellness is important for all uniformed personnel. In many departments, some individuals may gravitate to job tasks other than fire fighting because of personal necessity or interest. Such tasks may include EMS activities, rescue, hazardous materials response, or fire investigations. All such tasks, however, include significant physical and emotional stresses. Moreover, for all uniformed personnel (regardless of job assignment) the key test is the ability to perform active fire fighting.

What Does the Wellness Program Include?

The Fire Service Joint Labor Management Wellness-Fitness Initiative has five main components, with each main component presented as a separate chapter:

- Medical
- Fitness
- Medical/Fitness/Injury Rehabilitation
- Behavioral Health
- Data Collection and Reporting

Index

About the Editor

Stephen N. Foley serves as a Senior Fire Service Specialist with NFPA, with responsibility integral to fire fighter health and safety and fire service organization and deployment. In addition, he responds as a member of NFPA's Fire Investigation Unit and serves as Chair of the standards development and coordination subcommittee in the CBRNE arena for the Inter-Agency Board for Equipment Standardization and Interoperability.

Prior to working for NFPA, Mr. Foley served in various capacities within the municipal fire service, spending the last twelve years of his career as Fire Chief in Longmeadow, Massachusetts.

He has both undergraduate and graduate degrees as well as completion of the Executive Fire Officer Program in conjunction with the Kennedy School of Government at Harvard University. Mr. Foley has edited and authored numerous texts related to the fields of public fire protection. He continues to instruct and lecture both in the United States and abroad.